Paul Martin

studied biology at Cambridge University, where he acquired a First in Natural Sciences and a PhD in behavioural biology; and at Stanford University, where he was a Harkness Fellow and Postdoctoral Fellow in the School of Medicine. He subsequently lectured and researched at Cambridge University, and was elected a Fellow of Wolfson College. He is co-author, with Patrick Bateson, of *Measuring Behaviour* (1986; 2nd edn, 1993). He is married with a young family.

The Sickening Mind was highly commended in the Popular Medicine category of the BMA Book of the Year awards 1998.

Further praise for *The Sickening Mind*:

'Martin argues that if we are to remain healthy, our brains and our immune systems have to work in harmony, maintaining a synergistic relationship. Naturally this book will appeal to the hypochondriacs among us, but it has a lot more to offer than mere fodder for paranoia.' MICHAEL WHITE, *Mail on Sunday*

'[Martin] writes very clearly, and explains scientific concepts in such a way that anyone can understand them. One of the charms of this book is the author's use of literary examples to illustrate his points: for imaginative writers have instinctively known what plodding researchers with their test tubes and questionnaires have afterwards established on a firmer basis . . . this is a book that will instruct and delight people of sound constitution: but I should keep it firmly out of the hands of hypochondriacs.'

ANTHONY DANIELS, *Sunday Telegraph*

'Martin speaks directly and confidently to a reader presumed to be interested in his own life . . . [he] explains, extremely persuasively, what may be the role of the brain in our becoming ill. Most of us are stuck with an obstinate tendency to distinguish between mental and physical entities, the intangible and the tangible. This gives psychosoma … could read this book and adopt th … natic.'

…K, *Observer*

'Martin is a sure guide in this controversial field – and an eloquent one . . . He bases his case in part on the observations of Shakespeare and other literary giants of the past. But it is contemporary science which most strongly supports his contention that the relationship of mind to health is mediated both by our behaviour, and by biological connections between the brain and the immune system.' BERNARD DIXON, *Independent*

'The style is informal and direct, with a quirky humour . . . although this is a scientific book, it is accessible. Paul Martin argues brilliantly that all illnesses have psychological and emotional consequences as well as causes. I stand convinced.'
DARIANNE PICTET, *Literary Review*

'*The Sickening Mind* contains much that is firmly researched, thought-provoking, well argued and, above all, helpful as a focus of public debate.' NICK THOMAS, *Catholic Herald*

'Paul Martin is an exceptional writer, a biologist who not only writes clearly and interestingly about his own subjects, but also is widely read in literature and history and uses illustrations from many sources. He chooses good scientific and literary examples to fit the argument. This is a very well-written book . . . *The Sickening Mind* is a remarkable achievement. I have added it to my reference programme for use in academic work and recommended it to nonscientific friends as a book that, in the very best way, popularises science.' RICHARD MAYOU, *THES*

PAUL MARTIN

The Sickening Mind

Brain, Behaviour, Immunity and Disease

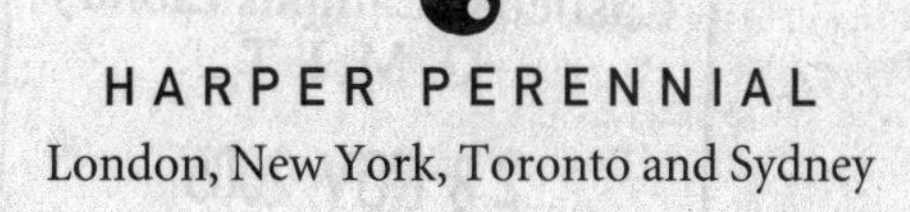

HARPER PERENNIAL
London, New York, Toronto and Sydney

Harper Perennial
An imprint of HarperCollins*Publishers*
77–85 Fulham Palace Road
Hammersmith, London, W6 8JB

www.harperperennial.co.uk

This edition published by Harper Perennial 2005

5

Previously published in paperback, with line corrections,
by Flamingo 1998 (reprinted three times)

First published in the UK by HarperCollins*Publishers*, 1997

ISBN-13 978-0-00-655022-8
ISBN-10 0-00-655022-3

The author and publishers are grateful to the following for permission to reproduce material: International Music Publications Limited, for 'Sex and Drugs and Rock and Roll' by Ian Drury and Chas Jankel, © 1977 Temple Mill Music Ltd, Warner/Chappell Music Ltd; Hall, J.G., 'Emotion and immunity' *Lancet*, 2, 326–327, © The Lancet Ltd., (1985); extracts from *Riceyman Steps* by Arnold Bennett by permission A. P. Watt Ltd., on behalf of Mme V. Eldin; extracts from *The Man with the Golden Gun* by Ian Fleming by permission Glidrose Publications Ltd, © Glidrose Productions Ltd 1965; extract from *Three Men in a Boat* by Jerome K. Jerome by permission A. P. Watt Ltd., on behalf of The Society of Authors Ltd.; extracts from *Down and Out in Paris and London* by George Orwell by permission A. M. Heath & Co. Ltd, copyright © The Estate of the late Sonia Brownell Orwell and Martin Secker and Warburg Ltd; extract from *Death of a Salesman* by Arthur Miller by permission Reed Publishers, © Arthur Miller 1949; extract from 'Do not go gentle into that good night' from *The Poems* by Dylan Thomas by permission David Higham Associates: originally published by J. M. Dent; Thanks also to Harriet Wasserman Literary Agency Inc., for permission to quote from *Seize the Day* by Saul Bellow; Macmillan General Books, for *Jude the Obscure* and *The Mayor of Casterbridge* by Thomas Hardy. Every reasonable effort has been made to contact copyright holders for all the extracts reproduced in this volume. However, it has not been possible to make contact with all copyright holders. The author and publishers would ask, therefore, that any copyright holder who feels a quotation contained herein may contravene their copyright contact HarperCollins*Publishers* at the address above.

Set in Postscript Minion

Printed and bound in Great Britain by Clays Ltd, St Ives plc

CONTENTS

ACKNOWLEDGMENTS

I am very grateful to Gillon Aitken, Bryn Caless, Paul Davison, Jonathan Evans, Kate Fallon, Joe Herbert, Philip Gwyn Jones, Alan Judd, Barry Keverne, Sally Kidner, Harriet Martin, James Serpell, John Sants, Michael Sharpe, Cecilia Thompson and Martin Trick for their encouragement and advice. The mistakes are all mine, however.

1

The Body of Knowledge

Most of the time we think we're sick, it's all in the mind.

Thomas Wolfe, *Look Homeward, Angel* (1929)

It is time to acknowledge that our belief in disease as a direct reflection of mental state is largely folklore.

Editorial, *New England Journal of Medicine* (1985)

Opening shots

You, dear reader, are going to die. Not for a long time, I hope, and painlessly. But die you undoubtedly will. And unless you die in the near future, and from unnatural causes, you will be ill before you die – probably several times. Some remarkable scientific discoveries have shown that your mind will affect your susceptibility to those illnesses and may have a substantial bearing on the nature and timing of your eventual death.

This book is about these scientific discoveries. It explores the ways in which your psychological and emotional state influence your physical health and how, in turn, your physical state affects your mind. It seeks to explain some of the extraordinary things that scientists have discovered in recent years about the interconnections between the brain, behaviour, immunity and health. By unravelling the biological mechanisms that underlie these phenomena, scientists can at last reconcile many commonplace notions about mental influences on health with a modern understanding of how the brain and behaviour affect the functioning of the body.

This is not intended to be a self-help book and I shall not be setting out detailed prescriptions for instant health or miracle cures for AIDS. The rapidly growing corpus of scientific knowledge about mind–body interactions has numerous potentially valuable applications in medicine, and I shall describe them. But practical action must be built on solid foundations of knowledge and understanding. As Sir Francis Bacon once remarked, 'Knowledge itself is power.'

Bacon also remarked that 'all knowledge and wonder (which is the seed of knowledge) is an impression of pleasure in itself.' I hope you will find the discoveries described here intriguing and worthwhile in their own right, regardless of their utilitarian value. We neglect the sheer wonder of scientific knowledge at our peril. Practical applications matter a great deal, but they are not the only fruits of science.

Let us get down to business by conducting a simple thought experiment. When you have read this paragraph shut your eyes and cast your mind back to the most mortifyingly embarrassing moment in your life, the worst that you can dredge up from the dank recesses of your memory. Think hard and choose the most awful, squirm-inducing calamity. Be brutally honest. Perhaps you committed an appalling social blunder at an august gathering, or said exactly the wrong thing at the wrong time. Close your eyes and re-live the incident in all its ghastliness, focusing on your own humiliation.

Have you blushed? Are your cheeks burning with embarrassment? If so, you have just demonstrated a mundane example of an important biological principle: that mere thoughts and emotions can generate very real physical reactions.

If you would like to demonstrate the empirical truth of this principle again, but in a different and more recreational way, close your eyes and conjure up your most arousing and succulent sexual fantasy. You surely must have one. Sit back and let your mind savour the luscious details of whatever erotic images it has chosen. Let the moist, quivering images run rampant. The physical consequences of what is now going on in your mind should, with any luck, be more fun than a blush.

The mind's influence on the body is usually more serious than a blush or a sexual *frisson*, however. It can even determine when we die. As an appetizer we shall consider two examples.

Iraqi SCUDs and Chinese grandmothers

During the Gulf War of 1991 Iraq launched a series of missile attacks against Israel. Many Israeli civilians died as a result of these attacks. But the vast majority of them did not die from any direct physical effects of the missiles. They died from heart failure brought on by the fear, anxiety and stress associated with the bombardment. They died because of what was going on in their minds.

How do we know this? After the war was over Israeli scientists analysed the official mortality statistics and found something remarkable. There had been a large and anomalous jump in the death rate among Israeli citizens on one particular day: the day of the first Iraqi attacks.

In the early hours of 18 January 1991 Iraq launched the first of several SCUD missile attacks against Israeli cities. Measured in terms of physical destruction, the Iraqi weapons were surprisingly ineffective. There were no deaths through physical injuries in the first attack and only two people were killed by the direct *physical* effects of SCUD detonations during the subsequent sixteen days on which missiles fell. And yet, on the day of the first attack, the death rate in Israel leapt by 58 per cent. A total of 147 deaths were reported, 54 more than would have been expected on the basis of previous mortality figures for that time of year. In statistical terms this was a highly significant increase; the odds against it arising from random fluctuations alone were enormous. What happened?

The evidence consistently pointed towards one conclusion: the sharp rise in death rate on 18 January 1991 was primarily a consequence of severe emotional stress brought on by fear of the Iraqi bombardment. It was the psychological impact of the SCUD missiles, not their physical impact, that claimed the majority of the victims.

The reasoning behind this conclusion was straightforward. The unexpected 'extra' deaths resulted in the main from heart failure or other forms of abrupt cardiovascular catastrophe. There was no increase in deaths from lingering diseases such as cancer, for example. As we shall see later, psychological stress can induce physiological changes which may prove fatal to someone who already has a diseased heart or clogged coronary arteries.

The 'extra' deaths were concentrated in areas of Israel where the levels of fear and anxiety were highest: regions that were known to be threatened by Iraqi missiles. In parts of Israel where SCUDs were not expected to land the mortality rate remained much the same as usual. Psychological studies carried out during this period indicated that the most stressful time for Israeli citizens was the few days leading up to the outbreak of war on 17 January, peaking on the day of the first SCUD missile attacks. There was enormous and well-founded concern about possible Iraqi use of chemical and biological weapons. The entire Israeli population had been issued with gas masks and automatic atropine syringes in case of chemical attack, and every household had been told to prepare a sealed room.[1]

After the first Iraqi strike had occurred, and turned out to be less cataclysmic than feared, there was a marked decline in levels of stress. As in other wars, the populace adapted to the situation with surprising speed. Then, as the fear and anxiety subsided, so the death rate also began to decline. There were seventeen further Iraqi missile attacks over the following weeks, but Israeli mortality figures over this period were no higher than average.

There is little doubt that many of the Israelis who died in the opening days of the war were killed by the Iraqi missiles. But there is equally little doubt that many of them died because of what was going on in their minds and not from physical injuries. Of course, their mental state was not the only thing that contributed to their deaths. Most, if not all, of those who died also had a pre-existing medical problem which made them especially vulnerable to the damaging effects of psychological stress. Many died because they had pre-existing coronary heart disease and their hearts gave out under the stress.

One of the recurring themes in this book is the simple point that disease and death seldom have single causes. The mind can help to precipitate illness or death, but this does not mean that bacteria, viruses, cancer cells, clogged arteries and other all-too-solid agents of disease are not also involved.

Mortality statistics have revealed another fascinating phenomenon.

Psychological factors can not only hasten death, as happened during the Gulf War, they can also postpone it. There is less likelihood of a person dying on the eve of an occasion that has symbolic significance for them, such as an important religious festival or birthday. There is compelling evidence that individuals on the verge of death can postpone their death for a few days until the special occasion has passed.

A clear demonstration of this phenomenon came from some scrupulously designed research conducted by David Phillips and Daniel Smith of the University of California at San Diego. They analysed the mortality statistics for Chinese people living in California to see whether there were any fluctuations in the risk of dying at around the time of the Harvest Moon Festival – an occasion which is of symbolic importance to Chinese people but not others.

Phillips and Smith found a large and statistically significant dip in the number of Chinese dying from natural causes just before the Harvest Moon Festival. This was followed by a corresponding and compensatory rise in mortality just after the festival was over. In the week preceding the festival the death rate among Chinese Californians was 35 per cent below the expected level, while in the week after the festival it was 35 per cent higher than expected. There was no overall change in the number of people dying, but some deaths that would otherwise have occurred just before the festival were somehow postponed until after it was over.

There is little doubt that this strange phenomenon of delayed death was specifically linked to the symbolic occasion of the Harvest Moon Festival. The dip and rise in the risk of dying was most evident among elderly Chinese women, who play a central role in the ceremonies. The Harvest Moon Festival is a movable feast – the date varies somewhat from year to year – so the fluctuation in mortality rate was definitely linked to the occasion itself, rather than to any specific calendar date. Furthermore, there were no comparable fluctuations in mortality among Jews and other non-Chinese Californians for whom the Harvest Moon Festival has no symbolic importance.

The analysis only looked at deaths from natural causes, so the phenomenon could not be explained by changes in people's propensity to commit suicide. Conceivably, some deaths might have been delayed because sick individuals took better care of themselves in the run-up to the festival, or because they received extra attention from their family and doctor. But the sheer scale of the phenomenon implied that something more profound was going on as well. In fact, the biggest fluctuations were in deaths caused by disorders of the heart and circulatory system, especially strokes and heart attacks. These are notoriously susceptible to psychological and emotional influences.

An almost identical dip and rise in mortality rate occurs among Jewish people around the festival of Passover. Like the Harvest Moon Festival, Passover is of cultural significance for one section of the community only and its dates vary from year to year.

The statistics reveal that the number of Jewish people dying from natural causes dips sharply just before Passover and bounces back with a compensatory increase immediately afterwards. Again, the fluctuation relates primarily to strokes and heart attacks and no such variation in mortality occurs among non-Jews for whom Passover has no personal significance.

Evidence like this strongly implies the existence of links between our mental or emotional state and our physical health. It is the scientific nature of these mind–body links, and their many ramifications, that we shall be exploring in this book.

Roundheads and Cavaliers

> All scientists know of colleagues whose minds are so well equipped with the means of refutation that no new idea has the temerity to seek admittance. Their contribution to science is accordingly very small.
>
> Peter Medawar, *A Note on 'The Scientific Method'* (1949)

> I am too much of a sceptic to deny the possibility of anything.
>
> T. H. Huxley, letter to Herbert Spencer (1886)

Contemporary attitudes towards the relationships between mind, body and disease are strangely confused. On the one hand we have the uncritical acceptance by the public, popular media and gurus of New Age medicine that the mind is both the source and the remedy for the majority of bodily ills. Set against these Cavaliers of mind–body interactions we have the Roundhead sceptics, who either dismiss the connections between psychological factors and physical disease as pseudo-scientific wishful thinking, or else simply ignore them altogether.

The tenet that psychological factors play a role in causing or curing bodily diseases is, of course, an ancient one – far older than modern medicine. Throughout history people have held deep-seated beliefs in the power of the mind to influence physical health, and down the centuries (until the twentieth century, anyway) physicians have explicitly linked physical wellbeing with mental wellbeing. It therefore comes as no great surprise to us if a major emotional upset such as bereavement, depression, divorce or redundancy later manifests itself in physical form. Our everyday experience, let alone statistical data from the Gulf War, seems to support this view.

But is this age-old notion of the mind affecting physical health a self-evident truth or merely unsubstantiated pseudo-science? Is it true that we are more likely to fall ill when we are stressed, anxious or depressed? Are individuals with certain personality types more susceptible to colds, allergies, heart disease or cancer? These are questions of profound medical significance. They are also fascinating scientific puzzles.

In ancient times healers worked on the pragmatic basis that the mind and the body are intertwined. Physical disorders could stem from problems in the mind and mental disorders could be reflections of bodily disease. Accordingly, physicians were encouraged to treat the soul and not just the body, using soothing words to comfort the patient's mind.

Ancient Greek medicine placed great emphasis on the curative power of *katharsis* – the purging and purification of the patient's soul. Plato and other great thinkers recognized that these psychological charms were remarkably effective in relieving physical ailments. They also recognized that these charms would not work properly unless both the patient and the physician believed in their curative powers.

Nowadays the supposedly damaging effects on health of anxiety, over-work, job insecurity and loneliness form a recurrent theme in the media, which preaches the message that stress makes us ill. The implicit connection between mental state and physical health seems to be uncritically accepted by an increasingly health-conscious public.

There has been an explosive growth in alternative and complementary forms of medicine, which tend to emphasize the underlying unity of mind and body. Around one third of the adult population has consulted a practitioner of the alternative medical arts at some time. Bookshop shelves groan under the weight of publications proclaiming the self-help gospel that health is all a matter of thinking the right thoughts and banishing negative emotions.

The self-help industry and New Age gurus offer us such tantalizing prospects as self-healing through love, thinking ourselves better from cancer, using the mind to heal all manner of dread diseases and, ultimately, reaching that holistic nirvana of health, happiness and self-fulfilment through the power of pure thought. It is easy to see why the sceptical Roundheads can be so dismissive of the mind–body Cavaliers.

A profound change in the pattern of diseases during the twentieth century may also have contributed to this trend. The infectious diseases that killed vast numbers until fifty years ago have almost disappeared from the wealthy industrialized nations – though not from poorer parts of the world. Their place in the league table has been taken by chronic degenerative disorders such as coronary heart disease and cancer. Diseases of the heart and circulatory system, cancer and accidental injuries now account for more than three-quarters of all deaths. In contrast, infectious and parasitic diseases account for less than 0.5 per cent of all deaths.[2]

The causal factors that contribute to these modern-day killers are much more complex than the relatively understandable causes of infectious diseases. We all recognize that tuberculosis is caused by bacteria, but cancer and heart disease are altogether more obscure. It is therefore easier to believe that the mind may play a role in their genesis. Factors as diverse as tobacco, red meat, slothfulness, insufficient fibre, childlessness, salt, pesticides, sunburn and radiation can cause serious diseases, so why not psychological stress or depression?

But is there any scientific basis for these beliefs? Just because people have always assumed something to be true does not make it so. After all, the earth was at one time assumed to be flat, stationary and at the centre of the universe. This belief appeared to be supported by everyday experience and was universally accepted as a self-evident truth. Yet it turned out to be completely wrong. Folklore, faith and dogma are not always reliable guides.

In stark contrast to the popular attitudes we have the inherent scepticism harboured by many scientists and doctors towards the notion that mere thoughts or emotions could possibly have an impact on such brutally physical processes as viral infections, coronary heart disease or cancer.

Scientific research in this field has often been tinged with a largely undeserved aura of crankiness. 'Psychosomatic' phenomena carry with them a whiff of self-indulgent fantasy, along with the implication that they lack both substance and scientific respectability. The suggestion that psychological and emotional factors play a causal role in disease is often regarded as an admission that the real (i.e., physical) origins of the disease are not yet understood. As Susan Sontag put it in her 1978 book *Illness as Metaphor*: 'Theories that diseases are caused by mental states and can be cured by will power are always an index of how much is not understood about the physical terrain of a disease.'

The belief in an intimate connection between mental state and physical health has had a decidedly rocky history in Western medicine, despite its promising beginnings in the civilizations of China and Greece more than two thousand years ago. By the end of the nineteenth century the overwhelmingly predominant approach to medicine was to focus exclusively on the disease and its identifiable physical causes, such as bacteria. Medical research could get to grips with bacteria, but thoughts and emotions were altogether too ethereal. The patient's mental state increasingly came to be seen as an embarrassing irrelevance – the province of psychologists and other faintly disreputable types rather than a proper concern of scientific medicine. In later chapters we shall consider why the mind and body came to be separated in Western thought, and how this estrangement of *psyche* from *soma* has had such an all-pervasive influence on modern science and medicine.

Yet even in the late nineteenth century there were notable exceptions to this rule. For instance, in 1884 Daniel Hack Tuke, one of the pioneers of British psychiatry, published the second edition of a work entitled *Illustrations of the Influence of the Mind Upon the Body in Health and Disease, Designed to Elucidate the Action of the Imagination*. In it, Tuke argued that the mind and body are inextricably linked through physiological processes; and that our mental state consequently affects our physical health and vice versa. State-of-the art research in the closing years of the twentieth century has come to much the same conclusion – and not before time.

History shows that important ideas can be ignored even if there is good evidence to support them. It is worth recalling the uncomfortable fact that compelling scientific evidence for the connection between smoking, disease and death was available for many years before it started to be taken seriously. Nowadays the link between smoking and all manner of dread diseases is almost universally accepted. Yet this was not always so. Scientists had suspected that smoking was bad for health long before the first solid evidence for a connection with lung cancer was published in 1950. During the 1950s and 1960s a succession of studies concluded that smoking increases the risks of lung cancer, heart disease and a host of other life-threatening conditions. Nevertheless, governments, the general public and even doctors remained sceptical of these links, and two decades passed before the research started to have an impact.

Contemporary physicians and scientists frequently dismiss the idea that the mind has a profound effect on physical health. To quote an editorial from a prestigious international medical journal: 'we have been too ready to accept the venerable belief that mental state is an important factor in the cause and cure of disease.' Another sceptic, also writing in a leading medical publication, comments that 'Mental stress is frequently blamed for the generation of organic disease, especially if it is of uncertain or complex aetiology, though without reliable or confirmatory argument . . . The morbidity of mental stress is commonly widely exaggerated.' Or consider this trenchant counterblast from a third scientific sceptic:

> During the last quarter of the 19th century many medical men asserted confidently that the stress of 'modern' life (i.e., all that gadding about in hansom cabs, paddle steamers, and railway trains) caused general paralysis of the insane [the final stages of syphilis]. Most of us now accept that this view was mistaken. I think that the notion that emotional factors have an important bearing on immunity, or on the cause or progress of cancer, comes into the same category.

In many respects the scientific evidence for connections between psychological factors and disease is stronger and more consistent than the evidence for certain other medical risk factors which are, nonetheless, regarded as less controversial. The putative links between dietary salt or cholesterol and heart disease are viewed with nothing like the same degree of suspicion and scepticism as psychological risk factors. Yet the scientific evidence that excessive salt or cholesterol in the diet actually cause heart disease in normal people is by no means conclusive. On the other hand, the evidence that psychological factors contribute to heart disease is wide-ranging and convincing, as we shall see later. There is a curious double standard at work here.

The fact is that most people, doctors and scientists included, find it inherently easier to believe in the reality of apparently simple physical causes of disease (such as cholesterol, salt, bacteria or viruses) than to accept that mere thoughts or emotions can affect our health. Partly as a result of such sceptical attitudes, research into the connections between the brain, behaviour, immunity and disease has, until recently, been remarkably neglected by mainstream medicine and seldom explained properly to the general public.

Can the starkly contrasting views of the uncritical Cavaliers and the sceptical Roundheads be reconciled? What are we to think when faced with conflicting claims about the mind's role in disease?

As we shall see, the scientific truth is subtler than either of these two extreme views. It is also far richer and more exciting. It turns out that the folklore was in certain respects right, while the sceptics were wrong in their sniffily dismissive attitude. Research has uncovered an array of solid, compelling evidence that the mind does indeed play a part in a multitude of disease processes, ranging from commonplace bacterial and viral infections to heart disease and even cancer.

Some completely fictitious case histories

Before delving into the science, let us turn our attention temporarily to storytelling. If it is true that our mental state influences our physical health then this fundamental aspect of human nature ought to have been noticed and reflected in literature throughout the ages. What can we see in the mirror that fiction holds up to the human condition?

The links between the mind, emotions, behaviour, disease and death have indeed been reflected in the lives of fictional characters over the centuries. The writers describing these mind–body phenomena obviously had no conception of their biological basis, but that did not stop them noticing and portraying the connections. Throughout this book I shall be referring to literary illustrations of the links between psychological factors and disease. But first, let me spell out what these fictional case histories are intended to convey and, perhaps more importantly, what they are not intended to convey.

I shall use literary allusions because they help to convey complex scientific ideas in a recognizable form. Well-turned examples drawn from literature are more cogent and more entertaining than any medical case history, no matter how supposedly authentic it may be. They also demonstrate the antiquity and universality of many of the concepts that underlie current theories. However, by citing fictional characters or situations to illustrate scientific theories I am certainly not implying that they constitute hard evidence in support of those theories. Fine words drawn from the imaginations of long-dead authors are clearly not the same as scientific data.

The idea that physical decline can stem directly from mental and emotional decline is a familiar theme in literature. Fictional characters often die from unrequited love, grief, shame or fury. Emily Brontë's

Wuthering Heights, for example, is positively bulging with characters whose mental states lay waste their physical health. Death and disease run riot throughout the book. Let me remind you.

Believing his childhood sweetheart Catherine has spurned him, the tempestuous Heathcliff vanishes. Catherine, who does in fact love Heathcliff, is deeply upset and ridden with guilt. She consequently becomes mentally unstable and physically ill. Three years later Heathcliff returns. Catherine is torn between her love for Heathcliff and her love for Edgar Linton, whom she has meanwhile married. She breaks down under the mental pressure, shuts herself in her room and sinks into delirium. The vengeful Heathcliff subjects Catherine to an emotional battering which further weakens her health and she dies giving birth, a victim of psychological torment.

The bereaved Heathcliff determines to achieve his longed-for union with Catherine through his own death. He locks himself in a darkened room and wills himself to die. Four days later a servant enters the room to find his body. Heathcliff's mind has killed him, as surely as if he had been shot:

> Kenneth was perplexed to pronounce of what disorder the master had died. I concealed the fact of his having swallowed nothing for four days, fearing it might lead to trouble, and then, I am persuaded he did not abstain on purpose; it was the consequence of his strange illness, not the cause.

Death by shame is the tragic fate awaiting Madame de Tourvel in Choderlos de Laclos's *Les Liaisons Dangereuses*, that tale of sexual intrigue among the enormously rich, enormously idle and enormously depraved aristocrats of pre-Revolutionary eighteenth-century France.

The young, pious and austere Madame de Tourvel is a devoted wife. Nevertheless, she is ruthlessly seduced by a satanic libertine, the Vicomte de Valmont – a man who 'has spent his life bringing trouble, dishonour, and scandal into innocent families'. Valmont is egged on by the Marquise de Merteuil, his equally amoral former lover. Together they plot their seductions with the cold, unemotional precision of a military campaign.

Facing stiff resistance from the virtuous Madame de Tourvel,

Valmont eventually breaks down her defences by convincing her that he will die from emotional torment unless she surrenders herself to him. Unable to resist his wiles any longer, Madame de Tourvel succumbs and Valmont has his wicked way with her.

The awful truth is then revealed – Valmont has cruelly deceived Madame de Tourvel and does not love her at all. In a fit of anguish and shame she flees to a convent, locks herself away and announces that she will not leave until she is dead. Her health rapidly deteriorates:

> A burning fever, violent and almost continual delirium, an unquenchable thirst . . . The doctors say they are as yet unable to diagnose . . . As long as she is so deeply affected, I have scarcely any hope. The body is not easily restored to health when the spirit is so disturbed.

The wretched Madame de Tourvel dies, destroyed by her grief and shame. No physical agent, other than Valmont, has intervened.

Fate, however, wreaks its just revenge on the perfidious Vicomte de Valmont and Madame de Merteuil. Valmont is fatally injured in a duel. Soon afterwards, the correspondence of Valmont and Merteuil, detailing their devilish seductions, is revealed and becomes the topic of widespread gossip. Madame de Merteuil is publicly humiliated, and one day later is afflicted by a virulent attack of smallpox which leaves her horribly disfigured and blind in one eye.

Another victim of physical decline brought on by seduction and shame is the eponymous heroine of Samuel Richardson's eighteenth-century blockbuster *Clarissa*. (At over a million words, it is also the longest novel in the English language.) Clarissa, a young lady of refined sensibilities, is seduced and raped by a superficially charming but unscrupulous bounder. Naturally – for this is eighteenth-century England – the dishonoured Clarissa is rejected and ostracized by her family and society. As her mental state declines, so does her physical health. Clarissa retreats into solitude and dies from grief and shame. Once again, a character's psychological and emotional state has been the prime cause of her death.

There are of course innumerable other examples, all making similar points about the impact of emotions on health.[3] We shall encounter a number of these in later chapters, and I hope it will be apparent that the notions they portray do bear some relationship to the reality revealed by modern science. However, at the risk of repeating myself, let me repeat myself: the fantasies of novelists are not the same as hard scientific evidence. Fortunately, there is plenty of that as well.

Invisible worms

> O Rose, thou art sick.
> The invisible worm
> That flies in the night
> In the howling storm,
>
> Has found out thy bed
> Of crimson joy,
> And his dark secret love
> Does thy life destroy.
>
> William Blake, 'The Sick Rose',
> *Songs of Experience* (1794)

The sharp divide between those who proffer psychological explanations for diseases and those who reject such theories in favour of purely physical causes is reflected in attitudes towards two particular disorders: tuberculosis and chronic fatigue syndrome.

Sir Peter Medawar, the Nobel Prize-winning immunologist and virtuoso science writer, once described tuberculosis as 'an affliction in which a psychosomatic element is admitted even by those who contemptuously dismiss it in the context of any other ailment.' There is abundant evidence, dating back hundreds of years, that the course and progression of tuberculosis are influenced by the sufferer's mental state. The physical health of tuberculosis sufferers shows a tendency to deteriorate when they are subjected to severe stress or emotional upsets.

Someone who is infected with *Mycobacterium tuberculosis*, the bacterium that causes tuberculosis, develops a protective immune response which can hold the bacteria in check and prevent them from multiplying. The resulting stalemate between body and bacteria can mean that the disease will remain dormant for years. But if something happens to compromise or weaken the body's immune defences, the bacteria can run riot and cause a resurgence of disease.

The importance of psychological factors in tuberculosis was widely acknowledged until well into this century. As early as 1500 BC, Hindu scripts described a wasting disease – almost certainly tuberculosis – of which one of the main causes was said to be sadness. Hippocrates, Galen and other medical luminaries of ancient Greece taught that grief, anger and other strong emotions played a major role in tuberculosis.

Throughout the seventeenth, eighteenth and nineteenth centuries many eminent European physicians stated that the causes of tuberculosis (or *phthisis*, as it was then known) included mental states such as 'a long and grievous passion of the mind', 'a mournful disposition of the soul', 'ungratified desires', 'profound melancholy passions' or 'disappointment in love'. During the eighteenth and nineteenth centuries, students of tuberculosis refined the theory of a link between the disease and a specific set of psychological characteristics. This was formalized in the concept of the tuberculosis-prone personality, or *spes phthisica*. As we shall see later, modern science has revived the concept of disease-prone personality types with the discovery of certain intriguing associations between personality traits, heart disease and cancer.

More familiarly, we have the romantic nineteenth-century notion that consumption (as tuberculosis was then known) is caused – or, more accurately, exacerbated – by an artistic temperament. Consumptive artists or writers were believed to be consumptive primarily because of their excessive artistic and aesthetic emotions, which literally consumed them. Consumption came to be something of a status symbol among the chattering classes.

Countless noted writers, artists and musicians of the nineteenth and early twentieth centuries did indeed suffer from consumption. Emily Brontë is but one example. She died of tuberculosis, a bitterly disappointed woman, within a year of *Wuthering Heights* being published to withering reviews. Other creative victims of tuberculosis included John Keats, Frédéric Chopin, Robert Louis Stevenson, Stephen Crane, Katherine Mansfield, Robert Tressell, D. H. Lawrence, George Orwell and Franz Kafka. It is said that Kafka's health was weakened by his chronic unhappiness, his hypersensitive personality, his problematic relationship with his father and several unfortunate romantic tangles. He died in an Austrian sanatorium in 1924, aged forty-one.

On the other hand, tuberculosis accounted for more than one in five of all deaths in the nineteenth century and was still common in Europe until well into this century. On statistical grounds it is hardly surprising that at least some prominent names were included among its victims.

The fiction of the eighteenth and nineteenth centuries reflects this preoccupation with tuberculosis and the prevailing attitudes towards it. We can choose from a panoply of heroes and heroines who sink into tubercular decline in the aftermath of romantic tragedy. We have, for example, the ever popular tale of *La Dame aux Camélias*, Alexandre Dumas the younger's tear-jerker about the passionate but doomed love affair between Armand Duval and the enchanting courtesan Marguerite Gautier.

The young Duval meets Marguerite and is captivated by her. They become lovers. Marguerite is already consumptive, but under the healing influence of Armand's love she abandons her dissolute lifestyle and her health improves. Meanwhile, Armand's well-meaning father secretly persuades Marguerite that if she truly loves Armand she will leave him for his own good. Marguerite demonstrates her love for Armand by doing just that.

The cynical Armand thinks Marguerite has abandoned him out of boredom so that she can resume her former life of late nights, promiscuity and wild self-indulgence. He therefore sets out to punish her by humiliating her at every opportunity and revelling in the spectacle of her suffering:

> ... for the last three weeks or so, I had not missed an opportunity to hurt Marguerite. It was making her ill . . . Marguerite had sent to ask for mercy, informing me that she no longer had either the emotional nor physical strength to endure what I was doing to her.

As a consequence of her deep unhappiness at their separation and the psychological battering she receives from Armand, the much-wronged Marguerite goes into terminal decline. Her consumption flares up and she expires. Armand realizes his error, but not until too late.[4] As the narrator sagely concludes: 'I have learned that one such woman, once in her life, experienced deep love, that she suffered for it and that she died of it.' Though Marguerite dies from tuberculosis, it is her emotions that have killed her. *La Dame aux Camélias* was, incidentally, modelled on Dumas' personal experience following his affair with the courtesan Marie Duplessis. Like the fictional Marguerite, Marie died of consumption not long after the affair ended.[5]

Until the early part of the twentieth century, popular medical literature was largely in tune with fiction in the way it emphasized the importance of psychological and emotional factors in tuberculosis. To quote one everyman's guide to medicine from the 1930s:

> Happiness is a mighty important factor in the treatment of tuberculosis . . . Mental brooding and loss of hope of recovery or of checking tuberculosis tends to drag the unfortunate individual into a deep chasm from which escape is rare.

So, what happened and why have attitudes changed?

What happened was that in 1882 the German scientist Robert Koch announced to the world that he had discovered the real cause of tuberculosis – the tubercle bacillus, *Mycobacterium tuberculosis*. Once it became known that tuberculosis was a bacterial infection, scientific interest in the role of psychological and emotional factors rapidly dwindled. The pendulum swung violently from the psychological to the physical. Open a contemporary medical text on tuberculosis and the chances are you will find little mention of psychology, emotions or the mind.

A lot of the old ideas about tuberculosis were plain wrong in their assumption that mental forces were sufficient to produce the symptoms by themselves; tuberculosis is undoubtedly a bacterial infection. Moreover, identifying a specific physical cause was immensely beneficial because it enabled medical science to find an effective remedy. Improvements in social conditions in the early twentieth century, followed by the introduction of effective antibiotics after the Second World War, led to an enormous decline in the incidence of tuberculosis in industrialized nations.

And yet it remains true that psychological and emotional factors do play a role in the disease. Later in this book we shall see how. It is not the nature of the tuberculosis that has changed, but the attitudes and interests of medical science.

CHRONIC FATIGUE SYNDROME

The misleading distinction between illnesses that are 'physical' (in other words, real) and illnesses that are 'psychological' (and therefore by implication not real) is starkly illuminated by the furore over chronic fatigue syndrome, otherwise known as myalgic encephalomyelitis (ME), post-viral fatigue syndrome or, if you read the tabloid press, yuppie 'flu.

It is with some trepidation that I thrust my head into the lion's den of controversy over the causes of chronic fatigue syndrome. Fierce arguments continue to rage and the medical establishment has yet to reach any consensus. In excess of eight hundred scientific publications have been devoted to the subject and the picture changes almost weekly. Those who suffer from the illness often have passionate views about its origins and anyone who gainsays them is asking for trouble.

The debate about chronic fatigue syndrome is relevant here because it exemplifies the false dichotomy between 'psychological' and 'physical' origins of illness. Throughout the controversy runs a seductively misleading vein: the implicit assumption that the illness must be either physical or psychological in origin. But first, what exactly is chronic fatigue syndrome?

Since 1988 the term chronic fatigue syndrome (CFS) has been used to describe a debilitating illness of unknown origin that has persisted for at least six months. As you probably know (for it is often in the news) CFS is characterized by a dreadful, disabling tiredness that is made worse by any physical exertion. This fatigue is accompanied by a motley assortment of other symptoms, including general malaise, intermittent fevers, pains in the joints, stiffness, night sweats, sore throats, poor co-ordination, visual problems, skin lesions and sleep disorders.

As if that were not enough, many CFS sufferers also experience psychological problems such as severe depression, forgetfulness, poor attention and lack of concentration. CFS can persist for years and it ruins the lives of those afflicted. Often they will be forced to give up work. Sufferers may show a measure of improvement over time, but the majority remain unwell for several years.

Cases of CFS have been reported in most industrialized nations including Britain, the USA, Canada, France, Spain, Israel and Australia. Sufferers tend to be young adults between twenty and fifty, though children can also be affected. According to the American Centers For Disease Control and Prevention, more than 80 per cent of CFS sufferers are women, most are white and their average age when the illness develops is thirty. Another common factor is that sufferers usually report having contracted some form of viral infection not long before the syndrome manifested itself.

As yet, no one has come up with a truly effective remedy for CFS. None of the drugs that have been used to treat the syndrome is of proven effectiveness and some may do more harm than good.

CFS, as currently defined, is a relatively recent phenomenon. (But then, so is AIDS; the fact that a disorder has only recently been recognized and defined does not detract from its reality.) Records of vaguely CFS-like syndromes, involving severe fatigue, muscle pains and other symptoms, date back at least two centuries. The medical history books, however, contain nothing that can be unequivocally compared with CFS before the second half of the nineteenth century, when neurasthenia became a common diagnosis. Incidentally, cultural stereotypes about the sort of person who was susceptible to neurasthenia were as strong in the nineteenth century as they are now about CFS. Neurasthenia was said to be a disease of affluent middle-class women, in much the same way that CFS has been inaccurately portrayed by the popular media as 'yuppie 'flu', a disease of affluent thirtysomething professionals.

It was not until the first half of the twentieth century that reports of a disorder corresponding to CFS started to accumulate. The first well-documented outbreak of a CFS-like disorder occurred in the 1930s in the USA and was attributed to a mystery virus. A similar mystery ailment afflicted the staff of a London hospital in 1955, in what became known as the Royal Free epidemic. The sufferers experienced persistent muscle pain and fatigue. To begin with the syndrome was referred to as benign myalgic encephalomyelitis. By 1956, however, it had proved to be anything but benign, and so it became known simply as myalgic encephalomyelitis, or ME.

Since they first appeared on the medical map, CFS-like illnesses have gone by a baffling variety of names including epidemic neuromyasthenia, neurasthenia, Iceland disease, Royal Free disease, atypical poliomyelitis, fibrositis, fibromyalgia, post-infectious neuromyasthenia, post-viral fatigue syndrome and myalgic encephalomyelitis. It is not certain that all these illnesses have been identical with what is now referred to as chronic fatigue syndrome. An analysis of twelve well-documented outbreaks of CFS-like disorders found they differed in various respects, notably with regard to neurological problems.

Now we come to the real meat of the problem. No one yet knows for certain what causes CFS. The arguments continue to rage and there are major divisions of opinion within the medical community. But what characterizes the whole debate – especially as it is portrayed in the popular media – is the implicit distinction between physical causes, which are held to be genuine, and psychological causes, which are held to be suspect.

With a few honourable exceptions, expert opinion on CFS divides neatly into two opposing camps. In one camp are those who maintain that CFS has a physical cause such as a virus or an immunological disorder. According to this view, the depression and other psychological symptoms that characterize CFS are consequences rather than causes of the underlying physical disorder.

In the opposing camp are those who argue instead that CFS is fundamentally a psychological disorder. According to this view, the physical symptoms such as exhaustion, muscle pains, fever and malaise, are manifestations of an underlying psychiatric problem.

Which view is correct? You may not be surprised to find that both are at least partially true. Many CFS sufferers have symptoms that match the diagnostic criteria for psychiatric disorders and organic disease. The evidence is undoubtedly complex and equivocal but it points towards one conclusion: that chronic fatigue syndrome has both physical and psychological components. Let us examine some of this evidence.

Most cases of CFS are preceded by a viral infection of one kind or another, and there have been repeated suggestions that a virus might lie at the root of the syndrome. For a long time the prime candidate was the Epstein-Barr virus, a member of the herpes virus family which is also responsible for glandular fever. During the 1980s chronic fatigue syndrome was widely referred to as 'chronic Epstein-Barr virus infection', as though its viral origins had been firmly established. Other candidates have included retroviruses (of which HIV is an example) and polio-like viruses called enteroviruses.

There is as yet no conclusive evidence to support the viral theory and it has therefore fallen out of favour. But even if viruses are not the prime cause of CFS, it remains highly plausible that a viral infection might help to trigger or precipitate the syndrome when other causal factors are also present.

Several other physical causes besides viruses have been proposed. One theory maintains that the primary symptoms of CFS are produced by hyperventilation – that is, abnormally rapid breathing. The evidence, however, is once again scant. Only a minority of CFS sufferers hyperventilate. On another tack, research at Johns Hopkins University in Baltimore has indicated that certain types of chronic fatigue (though not necessarily all cases of CFS) might result from abnormally low blood pressure. Yet another suggestion has been that CFS stems from a form of neurobiological disorder. One study revealed that more than a quarter of CFS patients had abnormal brain scans, and subtle changes have been found in the levels of neurotransmitter substances in the brain.

At present, the most favoured physical theories about the origins of CFS revolve around the immune system. There is growing support for the view that the symptoms of CFS result from a perturbation or abnormality in the sufferer's immune system. This immunological malfunction, it is argued, may be triggered by a viral infection which somehow throws the immune system out of kilter.

Evidence that CFS involves an immunological disorder is accumulating rapidly. Within the past few years various abnormalities have been found in the immune systems of CFS sufferers. These include alterations in the activity and surface structure of two important types of white blood cells: the natural killer cells and T-lymphocytes. (You will be hearing much more about these cells in later chapters.)

It is becoming increasingly evident that CFS is associated with, if not directly caused by, a persistent, low-level activation of the immune system.

If CFS really is an immunological disorder then why do some perfectly sensible scientists and physicians persist in regarding it as primarily a psychological disorder? They persist because there is highly respectable evidence to support their viewpoint as well.

Several of the symptoms associated with CFS are also seen in psychiatric illnesses, notably depressive and anxiety disorders. A substantial proportion of those who seek medical help for chronic fatigue turn out to have a recognizable psychological problem. The authoritative Centers For Disease Control and Prevention in the USA has concluded that approximately 45 per cent of all CFS sufferers have some form of identifiable psychiatric disorder *before* the onset of CFS. Researchers at the University of Connecticut School of Medicine found that as many as three out of four of the chronic fatigue cases they examined could be more easily explained by psychiatric problems such as depression. To add to the picture that the mind plays a central role in the illness, Australian researchers have discovered that CFS patients exhibit significantly more signs of hypochondria than other medical patients.

Psychological theories of CFS have tended to focus on depression. Over half of all CFS sufferers exhibit clear signs of clinical depression. Often the depression appears to have preceded the chronic fatigue, suggesting that it might be a cause rather than a consequence of the syndrome. Severe depression is usually accompanied by prolonged reductions in physical activity which could, in turn, lead to a debilitating decline in muscle function. People who lie in bed for long periods become physically weak. The sleep disturbances that typify some depressive disorders might also exacerbate the sufferer's fatigue. Furthermore, it is known that severe depressive disorders are associated with changes in the immune system.

But hold fast. It is equally clear that many CFS sufferers become depressed as a consequence of their illness. It is hardly surprising that those suffering from a debilitating but unexplained illness should become depressed and abnormally preoccupied with their health. Although more women than men suffer from CFS this should not be interpreted as evidence that CFS is primarily a psychological disorder, as a few sexist pundits have implied. There are several perfectly respectable organic diseases, such as rheumatoid arthritis, which show a marked preference for one sex over the other.

At present it is probably safe to conclude that the case for CFS being primarily a psychological disorder remains unproven. The evidence for some sort of immunological malfunction is too good to dismiss. There is, however, no doubt that CFS sufferers' psychological reactions to their illness do have an important bearing on their wellbeing and recovery. Whether depression is a cause or an effect of the syndrome, it becomes a major problem in its own right and can seriously impede recovery.

The controversy over CFS is further complicated by the attitudes of those who suffer from it. People who are afflicted by a serious and debilitating disorder such as CFS want their illness to be publicly recognized as having a medically respectable cause. For most people this means a physical cause, such as a virus or an immunological disorder, rather than a psychological cause. Any suggestion that their symptoms might result from a psychiatric problem tends to provoke outrage.

This attitude is understandable. Talk of psychological causes often carries with it an unjustifiable connotation that the illness is not quite genuine. There is usually a strong whiff of 'get a grip on yourself and snap out of it' in the air. Moreover, even in the late twentieth century there is still a wholly unreasonable stigma attached to mental illness. The average person would rather admit to having a physical illness, albeit a vague 'mystery' virus or obscure immunological malfunction, for this absolves them of any accusations of malingering, neuroticism or weakness of character. One unfortunate outcome of this desire for a physical explanation is the tendency, in some countries at least, for CFS sufferers to shop around until they find a physician who will give them the diagnosis they want.

Ironically, it turns out that the CFS sufferers who believe most strongly in a purely physical explanation have greater difficulty in recovering from their illness. This may be because they fail to confront and deal with the psychological problems that invariably accompany the illness.

Evidence to support this conclusion has come from a study conducted by Michael Sharpe and colleagues in Oxford. They found that a form of cognitive behavioural therapy, in which CFS sufferers were helped to re-evaluate their attitudes towards their illness, was of major benefit. More than 70 per cent of CFS sufferers who received the behavioural therapy regained their ability to function normally, compared with a success rate of 27 per cent for sufferers who received only standard medical care.

The pressure to attribute CFS to purely physical causes has also had a substantial influence on how the popular media deal with the subject. Newspaper and magazine articles, TV features and self-help books tend to emphasize physical explanations for CFS and neglect its psychological aspects.

A survey by researchers at the University of London found that 69 per cent of all articles on CFS which had appeared in national newspapers and women's magazines since 1980 had favoured physical causes, compared to a mere 31 per cent of research papers in scientific and medical journals. There appeared to be a systematic bias in the popular media towards reporting physical as opposed to psychological explanations. Even the choice of name was affected. Whereas scientific papers typically used the neutral term chronic fatigue syndrome, the popular media instead favoured the more medical-sounding myalgic encephalomyelitis (ME).

Similar attitudes apply to other illnesses which, like CFS, have been tarred with the psychosomatic brush. Asthma and allergies are familiar examples. So too are inflammatory bowel disorders such as Crohn's disease and ulcerative colitis. The pendulum of opinion has swung violently back and forth over the years. Half a century ago asthma was widely regarded as an essentially psychological illness. Nowadays it is normal to play down the role of psychological and emotional factors and instead focus almost exclusively on its immunological mechanisms and physical triggers, ranging from fitted carpets to car exhaust fumes. In truth, there are good grounds for believing that both immunological and psychological factors play important roles in these diseases. Nevertheless, the overwhelming tendency is to opt for one explanation to the exclusion of the other.

As we shall see in subsequent chapters, this centuries-old

as evidence of organic disease and for which medical help is sought' [my italics]. By this definition, the unfortunate victim might *feel* ill even though he or she has no underlying physical disease. In other words, mental state is the sole and sufficient cause of the physical symptoms. Such things do, of course, happen; we shall take a look at them in chapter 3. But they are not a major concern of this book. In fact, they are something of a distraction.

Psychological and emotional factors can determine whether or not someone becomes ill but they mostly do this by altering that person's susceptibility to disease. They are rarely the sole and sufficient cause of illness. A less misleading definition of 'psychosomatic' is one in which psychological factors play a contributing role in the development of the illness, alongside other factors such as bacteria, high blood pressure or smoking. But by this definition most illnesses in the Western world today can be termed psychosomatic.

The misleading conception of illnesses as mere phantoms, conjured up by the unconscious mind, has its roots in the psychoanalytic theories of Sigmund Freud. According to Freud and his disciples many mental and physical disorders have their roots in emotional conflicts, of which the patient may have no conscious awareness. These unconscious emotional conflicts are translated into physical symptoms such as pain, paralysis or loss of sensation. The symptoms are regarded by the sufferer – though not necessarily the rest of the world – as legitimate signs of a genuine organic illness. This dubious concept of psychosomatic illness lives on and can still be found lurking within the pages of popular health and self-help books.

Freudian psychoanalytic theories laid the foundations for what later became known as psychosomatic medicine, a field which came into being during the 1930s and 1940s. The earliest practitioners of psychosomatic medicine sought explanations for mysterious disorders such as asthma, allergies, arthritis, high blood pressure and peptic ulcers in underlying emotional conflicts and personality characteristics. Psychosomatic theories about asthma, for example, revolved around such notions as the fear of losing parental love. As a natural consequence of their Freudian leanings, many of the early psychosomatic practitioners tried to treat disorders like asthma and allergies using psychotherapy – with fairly mixed results.

We, on the other hand, shall be moving firmly within the realm of 'real' diseases like the common cold, herpes, coronary heart disease and cancer, rather than those shadowy and mysterious maladies to which the epithet psychosomatic is usually applied. The diseases we shall be focusing on in subsequent chapters are caused by real bacteria, real viruses, real clogged arteries or real cancer cells. They are most certainly not just 'all in the mind'.

Death, disaster and voodoo

Sometimes – quite often, in fact – people drop dead with little or no warning because something goes wrong with their heart. This phenomenon is called sudden cardiac death. It is normally defined as an unexpected heart failure within twenty-four hours of the first symptoms (if any) being noticed.

Sudden cardiac death accounts for about 15 per cent of all mortality from natural causes. Though victims may have no previous medical history of heart problems, autopsy generally reveals a pre-existing but hitherto undiscovered disease. Unfortunately, in more than half of all cases the first manifestation of this disease is death.

For centuries people have believed that severe psychological stress, grief, fear, anger or other strong emotions can trigger sudden cardiac death. There is massive anecdotal evidence that distressing events such as the death of a loved one, the loss of a job or even a heated argument can trigger a fatal heart attack. In recent years scientists have accrued a satisfyingly solid mountain of systematic evidence to confirm the anecdotes.

When scientists analyse the immediate precursors of sudden cardiac death they consistently find that a large proportion of its victims have experienced unusually high levels of emotional distress in the hours or days leading up to death. One study, for example, found that 40 per cent of men who died unexpectedly from heart failure had experienced a significant emotional upset, such as being involved in a car accident or receiving notification of divorce proceedings, within the twenty-four hours immediately preceding their death.

There have even been documented medical reports of individuals dying after being severely disturbed by upsetting thoughts or recollections of a traumatic experience.

One of the most common precursors of sudden cardiac death is the extreme fatigue and exhaustion known as burnout. Like consumption in the nineteenth century, burnout has become something of a bizarre status symbol. Burnout is seen as the 'red badge of courage' in professional circles, proof of Herculean labours and overwhelming workloads. (This says a great deal about present cultural values. In the nineteenth century consumption lent status because it supposedly denoted creativity and artistic passion; nowadays it is the sloggers we prize.)

Whatever the cultural overtones, there is a significantly higher risk of sudden cardiac death for victims of burnout. Those who exhibit the classic symptoms of intrusive anxiety, irritability and mental exhaustion may feel that way because of a mechanical fault in their heart. In many cases, however, burnout is more a symptom of prolonged psychological stress. In combination with a pre-existing weakness in the heart or coronary arteries it can easily be lethal. Dutch research which tracked the health of a large sample of middle-aged men over several years found that individuals who reported feeling mentally and physically exhausted at the end of the day were more than twice as likely to die from a heart attack. This was true even for men who had hitherto been free from any coronary heart disease.

In chapter 8 we shall be looking in greater depth at the biological mechanisms whereby the mind can damage the heart and coronary arteries. Suffice it here to say that there are plenty of well-understood biological mechanisms which enable stress-induced changes in the brain to trigger sudden cardiac death, especially where coronary heart disease is already present.

Sudden death can also be provoked by traumatic events on an impersonal scale. We have already considered the case of the Israeli citizens who died during the Gulf War from psychological stress generated by Iraqi missile attacks. Nature has conducted some of its own experiments in stress-induced death. Take earthquakes, for example. An analysis of mortality statistics immediately after a major earthquake will usually reveal a transient rise in the number of deaths from heart failure and other natural causes, unconnected with the direct physical effects of the earthquake. For instance, in 1978 the Greek city of Thessaloniki was hit by two earthquakes. Official records showed a marked increase in deaths from natural causes, especially heart failure. During the three-day period spanning the earthquakes and their immediate aftermath, the rate at which the local population were dying from heart disease shot up by 200 per cent and the death rate from other natural causes increased by 60 per cent.

Similarly, when Australian scientists investigated the aftermath of an earthquake which struck New South Wales in 1989 they found that the incidence of fatal heart attacks in the locality went up by 70 per cent. In these and other cases it was clear that psychological stress had brought about the premature deaths of vulnerable individuals.

Then we have those strange tales of voodoo, or 'hex', death. The unfortunate victim is ritually cursed by a witch doctor, voodoo priest, *bokor* or other symbolic authority figure. Once the death sentence has been pronounced the victim duly obliges by giving up the ghost and dying, usually within a few days. Competent and trustworthy authorities have been documenting instances of voodoo death since at least the sixteenth century, in places as far apart as Africa, South America, the Caribbean and Australia. It cannot be dismissed as the product of lurid fantasies.

The religious and cultural details vary, but reliable reports of voodoo death share certain basic features. First and foremost, the victim must be highly suggestible, with an unquestioning belief in the power of the sorcerer or witch doctor who curses him. He must also be totally convinced that he is powerless to do anything to save himself. An attitude of helplessness is essential: once the bone has been pointed or the curse uttered, the victim loses any will to live. Sceptics, scientists and tourists do not die from voodoo curses. A third important ingredient is social pressure. It speeds things along no end if everyone else in the victim's social world shares the same beliefs. Family and friends reinforce the victim's belief in the inevitability of death, abandoning the unfortunate individual to die in complete isolation.[1] The enormous importance of social relationships for mental and physical health is a theme we shall return to later.

Literature is replete with characters who drop dead from the effects of overpowering emotion. Shakespeare's King Lear, for example, dies of a broken heart when his favourite daughter Cordelia is cruelly murdered shortly after Lear is reconciled with her. On discovering Cordelia's body, Lear gives vent to his crushing grief:

> Howl, howl, howl, howl! O, you are men of stones!
> Had I your tongues and eyes, I'd use them so
> That heaven's vault should crack. She's gone for ever.

Then Lear drops down dead.

Trouble, strife and sickness

Intense emotion usually falls short of causing people to drop down dead; it may simply make them more vulnerable to illness. And here again, at least some of the folklore has withstood scientific scrutiny. Research has confirmed the existence of systematic links between psychological factors such as anxiety, stress, depression and hostility, and a wide range of physical disorders including minor infections, gut disorders, herpes, allergies, asthma, arthritis, coronary heart disease and cancer. Indeed, according to some characteristically controversial research by the London University psychologist Hans Eysenck, certain psychological measures of personality and behavioural style have a greater bearing on which individuals will die from cancer or heart disease over the following ten to fifteen years than whether or not they are smokers.

Anxiety and stress have frequently been linked with vulnerability to illness. Numerous long-term studies have found that people who experience pronounced feelings of tension or anxiety are substantially more likely to develop coronary artery disease, or die from it, over the following years. For example, an American study which tracked several hundred people over a twelve-year period found that individuals who exhibited high levels of psychological distress were roughly twice as likely to die as those with only average levels of distress. This connection between distress and death held up even when other medical risk factors such as old age, obesity, smoking, high blood cholesterol and high blood pressure were taken into account, so it was not merely a question of distressed subjects also being old, fat or smokers. Psychological distress was related to subsequent mortality in its own right.

Similar conclusions emerged from a Harvard University project. This investigated the health of former Harvard students whose psychological and biological profiles had been assessed thirty-five years earlier, as part of a series of laboratory experiments on stress. The way subjects reacted during the laboratory tests predicted their physical health years later. Individuals who displayed signs of severe anxiety during the original stress tests subsequently suffered from significantly more physical illnesses, including coronary heart disease, over the following decades. Responding anxiously to a stressful situation when a young adult proved to be a reliable marker for ill-health of all types in middle age. Another investigation by scientists at Harvard Medical School found that very shy children, who suffered from severe anxiety when in social situations, were more prone to allergic disorders such as hay fever.

It may help to look in greater detail at one specific example of a fairly subtle connection between psychological factors and subsequent disease. An American research project conducted in the 1970s investigated the psychological characteristics associated with infectious mononucleosis, otherwise known as glandular fever. This unpleasant and debilitating disease is prevalent among teenagers and young adults. The symptoms include a general malaise, fever, sore throat, loss of appetite, headaches, together with swelling of the lymph nodes or 'glands' in the neck, groin and armpits. Recovery can take many weeks. Occasionally, serious complications arise, such as damage to the liver or spleen.

The disease is caused by a type of herpes virus known as the Epstein-Barr virus (EBV), which we encountered in chapter 1 as a once-favoured cause for chronic fatigue syndrome. In common with other herpes viruses like herpes simplex (which causes cold sores and genital herpes), EBV can remain dormant in the body for years without causing any symptoms. Dormant viruses are normally held in check by the individual's immune system, but anything that weakens immunological control over the latent viruses can trigger the emergence of disease symptoms.

The subjects of this investigation comprised over 1300 young men entering the West Point military academy. On arrival at West Point each student was screened to see whether he was already infected with EBV. About two-thirds of the students carried the virus, which is typical for a normal population. The remaining third had not yet been infected. These potentially susceptible students were then tracked to see who would become infected with EBV. And here lies an important general point: not everyone who is exposed to disease-causing bacteria or viruses becomes infected. In fact, only about one in five of the originally virus-free students went on to be infected with EBV during their four years at West Point. Of those who did become infected, a quarter developed obvious clinical symptoms of disease. And here lies a second general point: not everyone who gets infected with disease-causing viruses or bacteria develops a clinical disease.

Psychological assessments revealed that those men who went on to be afflicted with infectious mononucleosis shared certain distinctive psychological characteristics. In particular, they tended to be the ones who had suffered most from academic pressure. Students who had the dispiriting combination of a strong motivation to do well, but a poor actual performance, had a greater likelihood of contracting infectious mononucleosis. And once they became ill these highly motivated but poorly performing students spent longer on average in hospital. They were more susceptible to the disease and when they got it, they got it worse.

Relatively minor traumatic events can also push up the odds of becoming ill. For example, Australian scientists found a marked increase in high blood pressure, gut disorders and diabetes among people who had been indirectly affected by a bushfire that occurred in southern Australia in 1983.

Long-term observations of normal families have shown that there is often an increase in family-related stress, or disruptive changes in family circumstances, in the period immediately before one or more family members develop infections. A number of studies of families in their home environments have unearthed associations between stressful conflicts and minor infectious illnesses such as coughs, colds, 'flu and sore throats. These stressful episodes tend to precede infections rather than follow them, implying that the stress contributes to the illness and not vice versa. In other words, it is not simply a matter of arguments arising because everyone is feeling ill and crotchety.

Research in the States has uncovered comparable links between stress and illness among children in rural Dominica. In the week following a high-stress event such as a big family upheaval, the risk of the children acquiring an infection of the upper respiratory tract increased by a factor of three.

Life events

For over thirty years scientists have been systematically exploring the idea that the risk of falling ill increases when we are exposed to a lot of disruptive changes or emotional turmoil. This research stemmed from the informal observations of certain perceptive doctors, who noticed that their patients often seemed to have experienced unusually large amounts of change and upset in the period before they fell ill. Further impetus came from a pioneering investigation of illness and absenteeism among the employees of the Bell Telephone Corporation in the 1950s. This indicated that employees with unsettled personal lives tended to suffer frequent bouts of illness and take more sick leave from work.

Suggestive observations such as these led psychologists to formulate the concept of life events. A life event is defined as any significant change in a person's circumstances which requires them to make psychological and practical readjustments. The disruptive event can be either desirable or undesirable; the prime criterion is that it causes a degree of upheaval.

Examples of life events include the death of a partner or family member, divorce, marriage, starting a new job, moving house or financial problems. At the other end of the scale, minor upheavals such as family holidays and Christmas are also classified as life events.

The basic hypothesis underlying this work is that any disruptive changes, whether desirable or undesirable, are potentially stressful and can increase our chances of falling ill.

Thousands of research projects have investigated the relationships between life events and health. The majority of these studies have used a standardized method for assessing life-event stress called the Social Readjustment Rating Scale. In its simplest form this involves asking each individual to record which of forty-three types of life event they have experienced over a specified period, usually between six months and two years.

Each type of life event is assigned a standard score according to its supposed severity, rated on a scale from 0 (least severe) to 100 (most severe). The maximum rating of 100 is awarded to the death of a spouse; divorce is rated 73; marriage, 50; changing to a different line of work, 36; moving house, 20; Christmas, 12; and so on. (Personally, I would rate Christmas at around 60, and anyone who has recently experienced the horrors of moving house may be excused for wondering at its modest rating.) A composite score is then calculated for each individual, taking account of both the total number of life events they have experienced and the relative awfulness of those life events. A high score can denote a few serious life events or a multitude of minor ones.[2]

If it is true that life events act as risk factors for illness then people who register high life-event scores should, on average, have more illnesses than those whose lives have been undisturbed by change. Simple. By and large, this is what the research has found.

A seminal early investigation looked at the effects of life-event stress on US Navy personnel during the Vietnam War. The results showed that individuals with the highest life-event scores suffered almost twice the number of illnesses over the following months as those with low scores. In another study scientists asked young men in a navy submarine training establishment to record the life events they had experienced over the previous twelve months; again, the incidence of life events correlated with subsequent illness.

The general conclusion from several thousand such studies is that people who have been exposed to lots of life-event stress have a slightly greater risk of illness. This increased risk applies across the board and seems to encompass virtually every form of ailment and disease under the sun, ranging from headaches, common colds, allergies and inflammation of the gums to mental illness, coronary heart disease, leukaemia, diabetes, tuberculosis and multiple sclerosis. Life-event stress also has an impact on childbirth; women who register high stress ratings during the year or so before pregnancy tend to give birth to babies with slightly lower birth weights and a slightly poorer overall state. Life events are even associated with an increased risk of minor accidents and sports injuries.

As well as suffering more episodes of illness, people with high life-event scores also tend to be ill for longer, have more severe symptoms and take longer to recover.

Not surprisingly, the adverse effects of life events are generally worse when the life events are severe, undesirable and clustered together in time. In the early days of life event research it was widely assumed that 'good' life events, such as getting married or starting a new job, were potentially just as damaging to health as 'bad' life events of comparable disruptiveness. However, more recent research has tended to support the common-sense assumption that, other things being equal, undesirable life events are inherently more damaging than desirable ones.

It has to be said that the link between life events and later illness is not as neat and simple as it sometimes appears. Some of the research on life events has been justifiably criticized for a variety of reasons. This is not the right place to debate the abstruse technicalities of research methodology. Nonetheless, the difficulties inherent in life event research are of broader relevance and therefore merit our attention.

First of all, the statistical correlation between life events and illness is highly consistent but it is also fairly weak. Life events do have a bearing on health, but not a very major bearing. Typically, life events account for only about 10–15 per cent of the total variation in the incidence of illness. A number of those who are exposed to stressful life events become ill, but most do not. Conversely, it is possible to fall ill despite living a life of unruffled stability. A phenomenon that is highly significant in a strictly statistical sense – meaning that the patterns in the data are more than just chance variations – may not necessarily be highly significant in a clinical or scientific sense.

A second fundamental point is that correlation is not the same as causation. The existence of a statistical association between two things is not proof that one of them causes the other. The population of the world and the age of the current pope are correlated, but there is no causal connection between the two. They both happen to be independently related to a third variable – time. So the correlation between life-event scores and illness does not by itself prove that life events are a direct cause of illness. The causation might even work the opposite way round; that is, chronic illness might conceivably precipitate life events. For instance, someone's marriage or career might run into problems because they are ill. And it may be the case that things which are classified as life events, such as sexual problems or changes in sleep patterns, could in fact be symptoms of an existing but undiagnosed illness.

In order to disentangle cause and effect in this type of research it is vital to establish which came first, the life events or the illness. There is plenty of evidence that life events do indeed tend to precede illness, which suggests that they may genuinely contribute to ill health.

A third pitfall with life event research, especially in its early days, has been its retrospective nature. When investigators ask subjects to recall their life events during, say, the previous year, great reliance is placed on frail and faulty memories. And therein lies a weakness. It is an awkward fact of life that most of us grossly over-estimate our ability to recall the past accurately and objectively. Ask any policeman, lawyer or judge about the reliability of witnesses to crimes. Psychologists have found that after a period of ten months people are typically able to recall life events with an accuracy of only 25 per cent. Conclusions that depend on people's memories of what happened to them one or two years ago are therefore bound to be suspect.

As well as the inherent difficulty of recalling past events accurately there is also a danger of systematic bias. People who are unwell may focus on a particular trauma in their past and assume it must have been responsible for their illness. We all have a basic need to find explanations for our illnesses and some people understandably attribute their poor health to traumatic experiences. But in doing so they inadvertently undermine the objectivity of the research data.

Fortunately, not all research on life events has had to rely on faulty memories. Instead, scientists have monitored groups of initially healthy subjects over a period of time, recording their life events and illnesses as and when they happen. This style of research is referred to as prospective, in contrast to the backward-looking retrospective method. And plenty of these prospective studies have borne out the link between life events and subsequent illness.

Another potential pitfall lies in failing to distinguish between an interviewee's actual health, as measured according to objective, clinical criteria, and what they say or think about their health. The problem here is not that people deliberately lie; the majority of those who volunteer to take part in scientific research try hard to be truthful. The real problem is that few of us are capable of being entirely objective about our own health. We all perceive and interpret our physical symptoms in different ways; something that would constitute a distressing malady for one person might not even be noticed by another.

Problems also arise if we attempt to measure health in terms of what is called sickness-related behaviour. This means behaviour like going to the doctor or taking sick leave from work. Sickness-related behaviour is obviously not the same thing as actual sickness.

The way humans respond when they think they are ill depends on other factors besides their state of health, including such mundane considerations as whether expert medical advice is freely and conveniently available. Sickness-related behaviour is often more a reflection of psychological factors than physical health.

To complicate matters further, people's perception of their own state of health varies according to their mental and emotional state. Anxious or stressed individuals, for example, are more apt to notice and worry about minor symptoms, interpret them as evidence of disease and seek expert help. Someone who has been experiencing lots of stressful life events is more likely to feel unwell and visit their doctor, but this does not necessarily mean that they are actually ill.

We shall be looking at this issue in more depth in the next chapter. Suffice it here to say that there is a world of difference between believing yourself to be ill, or going to the doctor, and having a clinically verifiable disease. For this reason, research that relies wholly on self-assessments of health or on sickness-related behaviour can be misleading. Such measures often say more about people's mental state than they do about their true physical health. I should add, however, that the dubious practice of using sickness-related behaviour as an ersatz measure of health is a pervasive problem in medical research and is certainly not unique to work on life events.

Despite these caveats there is consistent evidence, garnered from thousands of scientific studies, for a connection between life events and subsequent illness. It is now clear that even the mundane hassles of everyday life have an impact on physical health. Indeed, some scientists have argued that because these hassles are such a frequent occurrence their cumulative influence on health may be more pervasive than the effects of rarer, but more traumatic life events.

The general idea that psychological factors can affect susceptibility to physical illness is amply supported by research on other species. As in so many other respects there is nothing biologically unique about humans. Several decades of experimental work on other species have confirmed that various forms of psychological stress can increase (or, occasionally, decrease) animals' susceptibility to a wide spectrum of diseases, including bacterial and viral infections, heart disease and cancer.

For instance, when mice or rats are exposed to stressful situations, such as being physically restrained or subjected to unpleasant electric shocks, they become less resistant to infection with a whole range of bacteria, viruses and parasites including mycobacteria (the type of bacteria responsible for tuberculosis), herpes viruses, influenza viruses, polio viruses and the protozoa which cause toxoplasmosis. In one experiment, for example, frightening mice by exposing them to a cat significantly increased their vulnerability to infection with a parasitic tapeworm. (The cat was prevented from attacking the mice; the sight of it alone was enough to affect them.) Likewise, the social stress of being introduced into an unfamiliar flock makes chickens more susceptible to bacterial infections, while the stress of being transported renders cattle vulnerable to a form of viral pneumonia caused by the reactivation of latent herpes viruses.

The sheer volume of animal research in this field makes it impossible to describe more than a tiny and rather haphazard selection of examples. And some of the experiments, especially those performed in the dim and distant past, are too grisly and unethical to deserve a mention. We humans are not the only animals whose physical health can be damaged by upsetting events.

The mind and the common cold

The way in which psychological factors can affect our susceptibility to disease is illustrated by research on that most mundane of illnesses, the common cold.

For centuries it has been widely believed that stress makes us more prone to minor respiratory infections such as colds and 'flu. This has now been confirmed experimentally. It is surprising that until recently much of the scientific evidence regarding the effects of psychological factors on respiratory infections was suggestive rather than conclusive.

In one study, for example, researchers asked married couples to fill in a questionnaire each day for three months, recording the various stresses and hassles of everyday life together with their state of health. The results showed that respiratory infections tended to be preceded by a greater than average degree of stress. Typically, a few days before the onset of symptoms there would be a rise in the number of unpleasant life events and a drop in the number of desirable events.

Much firmer evidence came from a pioneering experiment in which psychologist Richard Totman and colleagues infected healthy volunteers with cold-inducing rhinoviruses, having first assessed each individual's psychological profile. It transpired that personality and previous exposure to stress had a significant bearing on both the risk of infection and the severity of the subsequent cold. Individuals with introverted personalities developed more severe colds, as did those who had experienced certain types of stressful life events.

The volunteers in this experiment were deliberately infected with viruses in order to avoid a potential ambiguity that had undermined previous research. Critics had pointed out that a correlation between psychological factors and colds could be attributed to varying degrees of exposure to cold viruses, rather than anything to do with biological resistance to infection. Individuals with shy personalities, say, or those who have recently experienced a traumatic life event, might be inclined to stay at home and would therefore have fewer opportunities to catch a cold.

By exposing all subjects equally to cold viruses Totman's experiment excluded this possibility. The fact that psychological measures still predicted the clinical outcome implied a more direct link between mental state and disease.

The technique of deliberately exposing people to bacteria or viruses in order to assess their vulnerability had, incidentally, been used before. In one hair-raising experiment in the early 1970s a group of healthy (and obviously well-motivated) volunteers were exposed to bacteria which cause a plague-like disease, with symptoms including prolonged fever, vomiting, headaches and swollen lymph nodes. Two days before they were infected each subject's stress level was assessed using standard psychological techniques. Those who registered the highest stress levels went on to have the most severe fevers.

Further compelling evidence for a connection between psychological stress and colds came a few years ago from a similar experiment. It is worth considering this experiment in detail because it illustrates some important general points.

Sheldon Cohen and colleagues recruited 420 healthy men and women and installed these worthy volunteers in residential accommodation at the British Medical Research Council's Common Cold Unit in Salisbury. They then used standard psychological techniques to assess the mental state and stress level of each volunteer. Specifically, the researchers noted the life events that each subject had experienced over the previous year; the extent to which subjects perceived themselves as unable to cope with the demands placed on them by life; and each individual's current emotional state. The volunteers were then exposed to a standard dose of cold viruses which matched the level of virus exposure one might expect to find in normal life. Each subject was given nasal drops containing one of five viruses capable of producing a common cold.[3]

Over the following week the subjects were monitored to see if they had been infected and, if so, whether they then developed clinical symptoms of a cold. Each day a doctor examined them for signs and symptoms of a cold using a standard checklist.[4] (So this experiment, you will notice, was immune from the criticism that stress might have affected the subjects' sickness-related behaviour as opposed to their actual health.)

The results of Cohen's experiment were clear and compelling. The more psychological stress an individual reported having been exposed to in the past, the greater their chances of infection with cold viruses and, once infected, the greater their chances of developing a clinical cold. Both the risk of viral infection and the risk of developing clinical symptoms increased in direct proportion to the amount of stress.

The correlation between stress, infection and illness was impressively strong. Individuals with the highest stress ratings were six times more likely to be infected with cold viruses than those with the least stress, and twice as likely to develop a cold. Moreover, these associations between stress, infection and illness held up even after the data had been adjusted statistically to remove any effects of other potentially relevant factors, including the subjects' age, sex, prior health, allergies, smoking and drinking habits, sleep and exercise patterns, diet, weight and education.

The technique of deliberately exposing the subjects to viruses ensured that they all had an equal opportunity to be infected. But you can be exposed to viruses without being infected. When you travel in a crowded train or bus you are regularly showered with exotic bacteria and viruses, but fortunately infection does not inevitably follow. Most of the time the bugs fail to make it past your skin or penetrate your inhospitable orifices. To establish that you have actually been infected it must be possible to recover viruses from your blood or body fluids, or show that your immune system has generated antibodies against the virus.

Exposure to viruses and subsequent infection are not the only steps along the path to illness, however. Not every infection develops into a clinical disease. The number of colds you will suffer in a lifetime represents a minuscule fraction of the number of cold virus infections you have had.

Detailed analysis of this experimental data enabled Cohen and his colleagues to tease apart the influences of stress on these two distinct components of disease. Whether or not someone was infected by the cold viruses depended primarily on how they were feeling at the time, especially their current perception of stress and negative emotions. But once they had been infected their chances of going on to develop a clinical cold depended more on their previous exposure to stressful life events than their current emotional state.

These results illustrate a general point: an individual's psychological state can exert different influences on the various steps in the pathway to disease, from initial exposure to disease-causing viruses or bacteria, through infection by those viruses or bacteria, to the development of disease symptoms and the behavioural response to those symptoms.

We have sampled some of the extensive evidence that what goes on in people's minds really does affect their chances of becoming ill or dying. The next question is how. It is time to consider the question of mechanism.

3

Psyche's Machine: The Inside Story

> Her pure and eloquent blood
> Spoke in her cheeks, and so distinctly wrought,
> That one might almost say, her body thought.
>
> John Donne, *Of the Progress of the Soul*,
> 'Second Anniversary' (1612)

By what means does the mind influence human susceptibility to disease? How can insubstantial thoughts or emotions produce a cold, let alone heart disease or cancer? After all, colds are caused by viruses not thoughts. We have seen evidence that our mental and physical states affect each other; what we need now is an explanation of how they do this. We need a mechanism.

In this chapter we shall explore the biological and psychological pathways by which the mind influences physical health – and, as we shall see, how physical health in turn influences the mind. This is the inside story of how the mind and body interact. There are three main strands to this story. First, our minds can make us *believe* we are ill, whether or not we really are ill in any objective, clinical sense. Our psychological and emotional state affects our perception of bodily symptoms and our reaction to those symptoms. This is the familiar (and generally misleading) connotation behind terms such as 'psychosomatic'. But the mind does more than influence our perception of physical wellbeing: it can genuinely affect our physical health. We come now to the second and third strands of the story.

The mind impinges on physical health in two fundamentally different ways: through our behaviour and, more directly, through our body chemistry. Psychological and emotional factors can lead us to behave in unhealthy or self-destructive ways which increase the risks of disease, injury or death. Smoking is an obvious example.

Meanwhile, beneath the surface, our mental state can alter our susceptibility to disease by influencing the body's biological defence mechanisms, most notably the immune system.

The perception of sickness

There is a fundamental distinction between illness – the sufferer's belief that something is wrong with them – and disease, which is a definable medical disorder that can be objectively identified according to agreed criteria. You can have a disease (such as early-stage cancer or coronary heart disease) yet not feel ill. Conversely, you can feel ill even though a doctor cannot detect any evidence of disease.

Many people who end up presenting themselves to a doctor have no identifiable organic disease. There is apparently nothing physically wrong with them. Yet they are still there in large numbers, claiming (and, in most cases, genuinely believing) that they are unwell. They are often referred to in rather loaded terms as 'the worried well'. But the majority of those who are suffering from vague, undiagnosed illnesses are not malingering. They really do feel ill and their ability to lead a normal life may be significantly impaired.

According to a report by the Royal College of Physicians and the Royal College of Psychiatrists, as many as half of all those who present themselves as out-patients for ostensibly medical reasons are suffering from psychological problems. Although they have physical symptoms such as pains, palpitations or breathlessness they have no detectable physical disease. Doctors perhaps understandably focus on the physical symptoms rather than the psychological problems. One consequence is that huge amounts of time and money are wasted on diagnostic tests and treatments for elusive diseases.

A substantial proportion of patients – a fifth or more – prove very difficult for doctors to deal with. Either their illness cannot be diagnosed at all, or, when a diagnosis is proposed, they find it unacceptable. Their treatment, if any, is frequently ineffective and they keep returning to the doctor over and over again, distressed and dissatisfied. These are the so-called heartsink patients. To make sense of what is going on we must once again turn to the mind.

Health and illness lie along a continuum. Often the dividing line between the two is arbitrary, and as much a reflection of our perceptions and expectations as it is of our true state of physical health. Our psychological and emotional state affects our sensitivity to bodily symptoms, our perception and interpretation of those symptoms and, finally, our propensity to seek medical help – whether or not those symptoms reflect a genuine disease.

Those who seek medical care do so because they have noticed certain symptoms, concluded that these symptoms constitute a real or potential illness, and decided to take action. Each of these steps is open to psychological and emotional influences. Individuals differ enormously in the extent to which they monitor their own health; in their willingness to put up with pain, discomfort and worry; and in their readiness to do something about it. The processes that culminate in a decision to visit the doctor depend on factors that are unique to each individual, including their social and financial circumstances, personality, experience, cultural background and genetic make-up. A lot can also depend on their current psychological and emotional state.

When a person is stressed or anxious they may become preoccupied with their health. There is a greater likelihood that they will notice (or imagine) physical symptoms; interpret those symptoms as indications of disease; and become sufficiently anxious about them to visit a doctor. They may also be more in need of the personal attention that they are perhaps not getting from others.

The heightened arousal that accompanies anxiety can make subtle bodily symptoms more noticeable. Moreover, the physiological changes that often accompany anxiety, such as headaches, churning guts or palpitations, may be interpreted as symptoms of disease. The mind can unconsciously create a medical mountain out of a molehill.

Our own perceptions are not the only ones that matter when it comes to assessing our state of health. The perceptions of those around us can also play an important role. Social pressures can reinforce, or even create, the perception that we are ill.

Imagine you are under a lot of stress. (Perhaps you don't have to imagine.) You have been told you are going to lose your job, your partner has left you and your personal finances are in meltdown. Unless you are exceptionally self-possessed your behaviour patterns will change noticeably. Perhaps you no longer relish the prospect of going for a drink with your friends; you feel depressed so you decline social invitations; you sleep badly and come to work looking tired; you are preoccupied with your problems and your performance accordingly suffers; you become irritable or keep bursting into tears; you go off your food and lose weight, or perhaps you turn to comfort feeding and pile on the calories instead.

Your friends and colleagues notice these changes and comment on them. They keep remarking that you don't look well; it must be the stress; perhaps you should see a doctor. Come to think of it, you don't feel too marvellous. Those headaches and the constant fatigue might be significant, and you *have* lost weight.

Before long you have convinced yourself that you are ill. You have certainly read enough magazine articles to know that stress is bad for your health. You take to your bed, or perhaps you trot off to see your doctor. To put it in the language of social psychology, social pressures have encouraged you to take on the 'sick role'. Now, you may indeed be genuinely ill; as we shall see, there is no doubt that stress can make us more susceptible to disease. But the thought processes that have led you to the conclusion that you are ill were driven largely by social pressure. Other people's minds, as well as your own, were involved in the process.

Consider, for example, the case of Colin Craven – the hypochondriac from hell in Frances Hodgson Burnett's children's classic *The Secret Garden.*

The obnoxious, bedridden Colin has been treated as an invalid, doomed to an early death, for all of his ten years. Everyone in Colin's orbit unquestioningly accepts that he is destined to be a crippled hunchback – that is, if he lives at all. They continually reinforce Colin's belief in his illness, reminding him of his weakness and urging him to rest. As one would expect, lying in bed all day has had a seriously debilitating effect on Colin's muscles; on the rare occasions when he does get up he feels genuinely feeble.

The egregious brat lies in bed all day with the family retainers pandering to his every whim. The servants live in fear of Colin's hysterical tantrums and dare not contradict him. The housekeeper privately recognizes that Colin is a victim of self-indulgence and hypochondria but would not dream of saying this to his face. To make matters worse, Colin's doctor is next in line to inherit the family property should Colin die and is therefore less than objective about the child's health. A London doctor who has had the temerity to suggest that Colin is not ill has been studiously ignored. Colin is immersed in his all-consuming hypochondria and sublimely unaware of how spoilt and unreasonable he is. Until his cousin Mary arrives.

Mary (who is not the nicest of children herself) rubbishes Colin's alleged medical condition during a fit of pique. She tells Colin bluntly that he has no trace of a lump on his back and is just being hysterical.

By challenging the unquestioned belief in Colin's illness, Mary has an electric effect on him. The supposed invalid soon comes to realize that there isn't anything wrong with him beyond his morbid state of mind. There is no lump on his back; he is thin and pallid because he refuses to eat properly; and he is weak because he lies in bed all day.

> So long as Colin shut himself up in his room and thought only of his fears and weakness and his detestation of people who looked at him and reflected hourly on humps and early death, he was a hysterical, half-crazy little hypochondriac who knew nothing of the sunshine and the spring, and also did not know that he could get well and stand upon his feet if he tried to do it. When new, beautiful thoughts began to push out the old, hideous ones, life began to come back to him, his blood ran healthily through his veins and strength poured into him like a flood.

With the help of cousin Mary, her rosy-cheeked proletarian chum Dickon and, of course, the Secret Garden, Colin is soon transformed into a 'laughable, loveable, healthy young human thing' who announces to the world that he is going to 'live for ever and ever and ever'.

A more delicate literary example of an indeterminate illness born of circumstance can be found in Tolstoy's *Anna Karenin*. Young Kitty

Shcherbatsky declines an offer of marriage from the worthy but unworldly Levin, expecting instead to receive a proposal from the dashing Count Vronsky. When Vronsky's anticipated proposal fails to materialize, Kitty, like a good nineteenth-century heroine, goes into a severe physical and mental decline which lasts for months. It is serious stuff and everyone is worried about the poor girl's health. Kitty's family doctor discusses her condition with a celebrated specialist whose help has been enlisted by the worried family:

> 'But of course you know that in these cases there is always some hidden moral and emotional factor', the family physician allowed himself to remark with a faint smile.
>
> 'Yes, that goes without saying', replied the celebrated specialist . . .

Kitty's family and friends are worried even though they are well aware that her condition has essentially psychological origins. Kitty is described as 'ill for love of a man who had slighted her.' Kitty's health does not improve and it is feared that she might actually die. Her anxious parents therefore take her on a foreign tour, where she encounters another young lady whose illness is also 'due to a love affair'. The passage of time and the distractions offered by foreign travel eventually bring about Kitty's recovery. Her illness and absence also allow circumstances to develop in her favour; she returns to Russia, marries the faithful Levin and (unlike the eponymous Anna) lives happily ever after.

Another way in which mental processes intrude into the domain of physical health is through the universal need for legitimacy. When we have decided that we are ill we want other people, and especially our doctor, to accept that we really are ill and not just malingering or being neurotic. Whether consciously or unconsciously, we want our putative disease to be accepted as genuine and not dismissed as a product of our fevered imagination. We need to legitimize our sickness by presenting the doctor with symptoms that will be accepted as evidence of a known organic disease. After all, no diagnosis means no treatment. As we saw in chapter 1, this can be a real problem for those suffering from poorly understood and controversial disorders such as chronic fatigue syndrome.

In his fascinating historical study *From Paralysis to Fatigue*, Edward Shorter has described how the physical symptoms that characterize so-called psychosomatic illnesses – those vague, undiagnosable ailments whose physical causes prove so elusive – have evolved over the years to keep pace with changing ideas about what constitutes a genuine disease. As society's perceptions and beliefs about disease have changed, so the symptoms of psychosomatic illness have also changed to keep pace with what is regarded as legitimate evidence of disease. Thus, in the eighteenth and nineteenth centuries it was common for people to succumb to hysterical paralysis, convulsions or 'fits of the vapours'. Paralysis of the legs was positively *de rigueur* among well-to-do young ladies of the nineteenth century. Nowadays, some would regard the symptoms of chronic fatigue and allergies as falling into the same category.

Shorter's historical analysis is interesting in that it demonstrates the powerful effect social pressures and cultural norms can have on patterns of symptoms. Actual diseases are another matter, however. There is nothing imaginary or unreal about many cases of chronic fatigue syndrome, allergies or other supposedly fashionable illnesses.

Our expectations also have an important influence on our perception of health. In industrialized societies like Britain and the USA general expectations of health have risen considerably in recent decades and continue to rise. As in so many other spheres of human activity, a consumerist attitude towards health has become the norm. People demand more in terms of their physical and mental wellbeing and are less willing to tolerate minor health problems which detract from their quality of life. That elusive – and probably illusory – gold standard of total health is increasingly demanded as of right even though, to quote one expert, 'deviance, clinically or epidemiologically defined, is normal'. This emphasis on positive health, as opposed to the mere absence of disease, is reflected in the explosion of interest in complementary or alternative medicine.

Huge advances in living conditions and medical knowledge have brought about large increases in life expectancy in many countries during the course of the twentieth century. Yet despite this we are apparently a sick bunch and getting sicker – if, that is, we define sickness in terms of perceptions and behaviour as opposed to objective measures of physical health.[1] Studies conducted in the USA in the late 1920s found an average of eight reported episodes of sickness for every ten people surveyed over a period of several months, whereas in the early 1980s the comparable figure was twenty-one sicknesses: an increase of 160 per cent. If we define sickness as seeking medical attention then the average person nowadays is 'sick' more than twice a year, compared with less than once a year in the 1920s. To be sick is normal.

Of course, what has increased over the decades is not the true incidence of diseases: it is our sensitivity to aches and pains; our tendency to ascribe them to physical diseases; our reluctance to put up with them; and our readiness to seek expert medical care.

Perish the thought, but just occasionally some of us have been known to concoct a tactical minor illness to get ourselves out of a predicament – perhaps as an excuse to avoid a dire social occasion or, less blatantly, to justify our poor performance in an exam, at work or in our personal relationships. Outright lying need not be involved. Gentle self-delusion is all that is needed. When sickness becomes an escape route from an unpleasant situation or embarrassment it is all too easy to convince ourselves that the symptoms are genuine. The 'sore throat' that conveniently gets the anxious child out of having to perform in the school concert can feel like a real sore throat.

Our minds, like Colin Craven's, can exaggerate the severity and significance of symptoms, causing us unnecessary distress and wasting doctors' time. But perceptions can shift in the opposite direction as well. An inert placebo 'drug' will often produce startling improvements in a patient's symptoms – provided the patient believes it to be a real medicine and expects it to have a beneficial effect. (We shall be revisiting the placebo effect later; it is yet another example of why the mind cannot be divorced from bodily health, even when we are dealing with apparently straightforward physical diseases.)

We all have the capacity unconsciously to blot out things we find too uncomfortable or upsetting to think about. This psychological defence mechanism is known as denial. However, the mind's ability to belittle or even ignore symptoms is something of a mixed blessing. Being excessively stoical or negligent about your own health is risky.

When people react to illness by denying the reality of their symptoms they may save themselves the unpleasantness of confronting an unpalatable reality. But their denial can be positively dangerous if it prevents them from seeking timely medical attention. A woman who fails to notice a lump in her breast, for example, or chooses to disregard it until her breast cancer is at an advanced stage, may pay for her insouciance with her life.

It is an unfortunate fact that people are less likely to seek medical help if it is difficult, inconvenient or embarrassing for them to do so – perhaps because they are too busy, or cannot afford the fees, or because they are simply afraid of calling a doctor out on a false alarm. Heart attacks are notoriously more likely to prove fatal at weekends, when it is inconvenient or potentially embarrassing to seek expert medical help. The lives of countless heart attack victims might have been saved had they not incorrectly attributed their chest pains to indigestion.

The disastrous consequences of denial are sombrely portrayed in Arnold Bennett's *Riceyman Steps*. The tightfisted Clerkenwell bookseller Henry Earlforward has cancer of the stomach but steadfastly denies that he is ill. Earlforward insists that it is merely a temporary indisposition and that he has a constitution of iron.

For a long time Earlforward's wife interprets his lack of interest in food as a symptom of his miserliness rather than any medical problem. Even when it becomes obvious that the emaciated bookseller is gravely ill he obstinately refuses to be examined by a doctor, let alone admitted into hospital. His wife rails at him for concealing from her the seriousness of his illness until it is too late to do anything about it. She tries hard to persuade Henry to accept medical help, but is forced to concede for 'nobody can keep on fighting a cushion for ever'. Faced with Henry's bland obstinacy, his wife and doctor eventually abandon their attempts to help him and he dies from his cancer – a victim of his own misplaced psychological defences.

Whether or not an illness has psychological origins it will certainly have psychological consequences. Feeling ill for any length of time is a psychologically debilitating experience. One of the simple but important ideas I hope to convey in this book is that the relationships between mind, body and disease work both ways. The mind affects the body and hence physical health. Conversely, physical health affects the mind and hence our thoughts, emotions and behaviour.

All but the most trivial of illnesses produce some sort of emotional reaction, whether it be mild irritation, anxiety, anger, denial or depression. Other things being equal, a serious illness should provoke a more intense emotional reaction than a minor illness. But other things seldom are equal. Illness means different things to different people, and just because an illness is not life-threatening this does not mean the sufferer will be emotionally untouched by it. An individual who has never before experienced any significant illness, pain or discomfort may be upset by relatively minor symptoms which would seem insignificant to someone who has suffered a string of serious diseases.

Our emotional responses to illness can have a crucial bearing on our recovery and future health. If being ill makes us depressed we may become careless about adhering to our doctor's advice or taking our medicine. This may, in turn, impede recovery. Whether or not a cancer patient adheres strictly to a programme of radiotherapy or chemotherapy can have a major impact on their chances of survival. There are patients who simply give up and sink into decline.

In extreme cases the emotional reaction to an illness can prove a bigger problem than the illness itself. Severe depression is far more debilitating and intrusive than many physical ailments. As we shall see in the next chapter, severe depression can also have detrimental effects on immune function and subsequent health, creating a spiral of decline. Doctors and patients ignore the psychological and emotional consequences of illness at their peril.

Finally, please do not go away with the impression that an individual's perception of their own health is an entirely meaningless or deceptive index, indicating only their degree of hypochondria. On the contrary. Research has shown that in certain respects perception is a good guide to reality. Although our subjective judgement is not always an accurate index of our current state of health, it does provide a reasonably good predictor of our long-term risk of dying prematurely. Depressing though it may be if you are an arch hypochondriac, the research indicates that people who believe they are unhealthy do die younger on average. Moreover, perceptions are clearly important for practical and economic reasons: people's perceptions of their health, rather than objective measures of health, are what largely determine their initial usage of medical facilities.

Bad behaviour

Sex and drugs and rock and roll
Is all my brain and body need
Ian Dury, 'Sex and Drugs and Rock and Roll' (1977)

A cousin of mine who was a casualty surgeon in Manhattan tells me that he and his colleagues had a one-word nickname for bikers: Donors. Rather chilling.
Stephen Fry, *Paperweight* (1992)

Our minds can have a profound impact on the physical health of our bodies by altering the way we behave. Psychological and emotional factors can dispose us to do all manner of unhealthy and self-destructive things. The self-destruction may be absolute and abrupt, as in suicide or fatal accidents, or gradual and cumulative, as in smoking.

Stress and anxiety, for example, can prevent us from sleeping properly and make us more inclined to smoke, drink excessive amounts of alcohol, eat too much of the wrong sorts of food, omit to take our medicine, neglect physical exercise, consume harmful recreational drugs, indulge in risky sexual behaviour, drive too fast without wearing a seat belt, have a violent accident, or even commit suicide (though not usually all at once).

Anna Karenin offers an impressive catalogue of self-destructive behaviour engendered by psychological and emotional trauma. Anna abandons her husband, the colourless bureaucrat Karenin, for the dynamic Count Vronsky. But their love is doomed and the emotional pressures on Anna build up to a fatal climax.

As a preamble to her eventual self-destruction, Anna nearly dies giving birth to Vronsky's illegitimate daughter. In what she thinks are her final hours Anna appears to reconcile herself with her husband. Mad with emotional torment at this turn of events, Vronsky goes off and shoots himself – but not fatally. Although Vronsky is an army officer, and therefore presumably capable of hitting his own heart at point blank range, the bullet misses. He is seriously wounded – enough to make it a meaningful parasuicidal gesture – but does not die. Anna and the baby go to live with Vronsky, but her husband refuses to divorce her and she becomes a social outcast. The strain of her position renders Anna increasingly unstable and she develops paranoid delusions about Vronsky's supposed unfaithfulness. Consumed by the madness of her passion, Anna suddenly decides that she must end her torment and punish Vronsky for his imagined misdeeds by killing herself. Anna famously ends her own life under the wheels of a train:

> 'There,' she said to herself, looking in the shadow of the trucks at the mixture of sand and coal dust which covered the sleepers. 'There, in the very middle, and I shall punish him and escape from them all and from myself.'

And she does. And there is more. Almost insane with grief at Anna's death, the bereaved Vronsky volunteers to fight, and very probably die, in a war between the Serbians and the Turks. Vronsky no longer places any value on his life and relishes the prospect of death: 'I am glad there is something for which I can lay down the life which is not simply useless but loathsome to me. Anyone's welcome to it . . .'

The melodrama of Anna Karenin's suicide and Vronsky's death wish are positively restrained in comparison with the high-camp posturings of Werther, the suicidal hero of Goethe's *The Sorrows of Young Werther*. This eighteenth-century piece of unfettered Teutonic sentimentality tells the tragic tale of an unbalanced youth who tops himself after a bad dose of unrequited love.

The story is a simple but eternal one. Werther loves Lotte. Oh, how he loves her! But, alas, he cannot have her. Lotte is already promised to the worthy Albert and soon marries him, leaving Werther to wallow in emotional excess. He sheds a thousand tears one moment and 'overflows with rapture' the next, and each step on the way is recounted in copious letters to his long-suffering chum Wilhelm. So it comes as no surprise that, denied his one true love, Werther decides to end it all. Characteristically, his suicidal decision is reached only after much beating of chest, gnashing of teeth, shedding of tears and general languishing in melancholy, during which time an unkind reader might be forgiven for urging the lad to get on with it. Even when Werther finally does get round to pulling the trigger he takes several hours to die.

Incidentally, the tragic tale of young Werther had a fairly profound effect on the health of a number of readers. So resonant was Goethe's writing with the romantic spirit of the times that the book triggered an epidemic of copy-cat suicides and was consequently banned in many places.[2]

All the leading causes of death in industrialized nations – including heart disease, cancer, accidental injury and AIDS – depend to some extent on how we behave. Smoking, eating habits, alcohol consumption, physical exercise, sleep patterns, sexual behaviour and choosing to wear a seatbelt, to name but a few, have ramifications for our health and wellbeing.

In industrialized societies, for example, accidental injuries and violence now account for at least half of all deaths among young men: a fact that is not wholly unrelated to the behavioural characteristics of young men. In extreme cases people who are very depressed or upset commit suicide or deliberately behave in a way which invites serious injury or death. Severe depression can lead to self-destructive behaviour. Besides making us act in positively unhealthy ways, psychological factors like anxiety, stress or depression can also inhibit us from engaging in activities that are beneficial to health, such as physical activity or social relationships with others.

In certain cases, such as crashing your car or committing suicide, the causal connection between behaviour and the subsequent damage to health is pretty obvious and requires no intimate knowledge of medical science to understand. Thanks to education and constant repetition in the media, less obvious connections between behaviour and health are also now widely recognized. The public accept that there are links between smoking and all manner of fatal diseases; between slothfulness and heart disease; between alcohol abuse and cirrhosis; and between unprotected sex and AIDS.

A stark illustration of how behaviour affects health is provided by

AIDS. There are enormous geographical variations in the incidence of HIV infection and AIDS. For example, the incidence of AIDS in Honduras is fourteen times higher than in neighbouring Guatemala. Even within a single country or a single city there are massive variations in rates of infection between different social groups.

Since the HIV retrovirus was discovered to be the causal agent for AIDS in 1983 it has become clear that these large variations result primarily from differences in people's behaviour – especially their sexual behaviour, which remains the route by which the virus is transmitted in the vast majority of HIV infections. It is generally accepted that a practical vaccine or cure for HIV/AIDS is at least a decade away.[3] In the meantime, the only effective means available for limiting its spread is to change the way we behave.

There are plenty of commonplace behaviour patterns that kill people gradually but in huge numbers. Smoking is the prime example. As long ago as 1604 King James I, in his treatise *A Counterblast to Tobacco*, did not exactly pull his punches when he described the new-fangled habit of smoking as:

> A custom loathsome to the eye, hateful to the nose, harmful to the brain, dangerous to the lungs, and in the black, stinking fume thereof, nearest resembling the horrible Stygian smoke of the pit that is bottomless.

Smoking is the riskiest thing that most people will ever do in their lives. At present, smoking-related diseases account for 15–20 per cent of all deaths and result in over 100,000 premature deaths every year in Britain alone. Smoking greatly increases the risk of lung cancer, now the commonest fatal cancer in Britain. Smokers are ten times more likely to die from lung cancer than non-smokers and around 90 per cent of lung cancers are attributable to smoking.

Smoking also increases the risks of various other fatal or debilitating diseases including coronary heart disease (the biggest cause of death in most industrialized countries), chronic bronchitis, emphysema, and cancers of the oesophagus, bladder and pancreas. A quarter of all deaths from coronary heart disease are smoking-related. As if that were not enough, smoking causes birth complications and doubles the risk of a pregnant woman miscarrying.

Think about these statistics from the British Medical Association. The average risk that you will die from leukaemia within the next year is about 1 in 12,500. The average risk that you will die in a vehicle accident is 1 in 8,000. If you are, say, forty years old, your risk of dying from natural causes of any sort during the next twelve months is 1 in 850. However, if you smoke ten cigarettes a day your odds of dying within the year are 1 in 200. Or look at it another way: take a random sample of a thousand young men who smoke; on the basis of actuarial data it can confidently be predicted that one of these young men will eventually be murdered, six will be killed on the roads and two hundred and fifty will die prematurely from the effects of smoking.

Smoking is clearly bound up with what goes on in people's minds. The reasons why individuals start smoking and why they then find it impossible to quit are neither simple nor well understood. Psychological studies of smokers have, however, confirmed the truth of several common assumptions.

It is indeed true that people who are depressed or stressed are more likely to smoke (and, consequently, more likely to die from lung cancer). Smokers really do experience a stronger desire to smoke at times of heightened anxiety. To add to their problems, psychological stress is associated with a higher failure rate among smokers trying to kick the habit. One long-term study of smokers found that individuals who had been depressed as much as nine years earlier were 40 per cent less likely to be successful in their attempts to give up smoking.

It gets worse. The psychological and emotional factors that make people inclined to smoke induce them to do other unhealthy things as well. Research has shown that moderate-to-heavy smokers are, on average, significantly less conscious of health-related issues, hold less favourable attitudes towards healthy behaviour and have a generally less healthy lifestyle in comparison with non-smokers or light smokers. (Conversely, wholesome behaviour patterns also come in clusters; researchers at Harvard University Medical School found that individuals who drank only decaffeinated coffee also tended to eat lots of vegetables, take regular exercise and wear their seatbelts.)

As well as prompting people to smoke, stress is also linked to increased alcohol consumption – at least, in certain types of individual. The health implications of excessive drinking can be profound. Approximately 20 per cent of all male in-patients in British hospitals have alcohol-related problems. Alcohol can rot people's livers and kill them in drunken accidents (though alcohol is not the only recreational drug capable of damaging health: there is reasonably good evidence, for example, that marijuana impairs the immune system, with potentially adverse consequences for the health of long-term users.)

The perils of the grape are amusingly described in *Othello*. The scheming Iago lures the unwitting Cassio into getting steamingly drunk, as a result of which Cassio lands himself in serious trouble and loses his job. On sobering up, Cassio bemoans the loss of his reputation and curses the demon drink:

> 'Drunk! And speak parrot! And squabble! Swagger! Swear! And discourse fustian with one's own shadow! O thou invisible spirit of wine, if thou hast no name to be known by, let us call thee devil! . . . O God, that men should put an enemy in their mouths to steal away their brains! That we should with joy, pleasance, revel and applause, transform ourselves into beasts!'

Literature is amply stocked with characters who drink themselves into an early grave in reaction to emotional crisis or unhappiness. There are roistering drunks who drink to escape boredom or poverty, like J. P. Donleavy's *Ginger Man*, Sebastian Dangerfield. There are determined drunks who drink to escape from grief. In *Wuthering Heights*, the unfortunate Hindley Earnshaw becomes a hopeless alcoholic after the death of his wife (from consumption, naturally) and drinks himself into the grave by the age of twenty-seven. And there are aimless drunks who drink to forget their own pointlessness. In F. Scott Fitzgerald's *The Beautiful and Damned*, for example, we have Anthony Patch, an independently wealthy and well-educated young man blighted by indolence, boredom and melancholy. A turbulent marriage and self-imposed idleness push him into self-destructive alcoholism and he degenerates into 'Anthony the poor in spirit, the weak and broken man with bloodshot eyes'.

Incidentally, when it comes to self-destruction by alcohol the track record of doctors is almost unrivalled. As a profession, they rank second only to pub-owners and bar staff in the league table of deaths from alcohol-related liver disease. Doctors are 3.4 times more likely than the average worker to die from cirrhosis of the liver. According to one 1995 estimate, as many as one in twelve British doctors is addicted to alcohol, drugs or both, thanks mainly to the enormous stress the majority of them are constantly under. (But I should not be too smug about this statistic because 'literary and artistic workers' also fare badly, with twice the average death rate from cirrhosis.)

On the other hand, moderate alcohol consumption can be an effective buffer against stress – and here again science has only of late managed to verify thousands of years' worth of everyday experience. Psychological studies have confirmed what countless millions of people have discovered for themselves, namely that when we are under stress we often feel less anxious if we drink alcohol. (A moderate intake of alcohol also appears to reduce the risk of coronary heart disease, but that is another story.) Sir Winston Churchill's opinion was clear: 'I have taken more out of alcohol than alcohol has taken out of me.'

There is nothing surprising about the fact that alcohol has its good side. It has, after all, been an intimate part of human life since the dawn of civilization. Alcohol was in use for medicinal purposes (in the literal rather than euphemistic sense) over four thousand years ago and was probably quaffed for recreational purposes long before that.

Opinions differ as to when exactly humans first discovered the joys of booze, but there is evidence that wine was being drunk in Transcaucasia eight thousand years ago – long before the wheel was invented. Some authorities have argued that Stone Age man was cultivating vines as early as ten thousand years ago. Wine growing was well established in the Middle East by 4000 BC and was an integral part of daily life in ancient Egypt and Mesopotamia. It says something that wine is mentioned 150 times in the Old Testament.

Then there are the social benefits of communal drinking to add to the purely pharmacological pleasures of alcohol. Samuel Johnson spoke for many when he declared that: 'There is nothing which has yet been contrived by man, by which so much happiness is produced as by a good tavern or inn.'

Yet the things that give us pleasure carry risks, and we are very poor at assessing those risks. While we consistently overestimate the dangers posed by rare or exotic threats like plane crashes, murders, nuclear accidents or shark attacks, we tend to disregard the risks of common killers like heart disease and vehicle accidents. We are especially prone to underestimating the risks arising from our own behaviour, such as smoking, travelling in cars, abusing alcohol or having unprotected sex.

Smokers now acknowledge the unappetizing fact that their behaviour significantly increases their risk of dying prematurely from heart disease or cancer. Nevertheless, psychological research has established that they seriously underestimate the magnitude of that risk. There is a consistent 'optimistic distortion' of perceived health risks among smokers; they know smoking is bad for them but they do not recognize just *how* bad. No matter how often the statistics are quoted they do not seem to sink in. One reason why the health consequences of smoking have such a muted impact on people's perceptions is the large delay, often measured in decades, between starting to smoke and falling ill.

If you should happen to be an overweight, tobacco-addicted, boozing, couch potato who loves fried food, you can take a few crumbs of comfort from the fact that others' attempts at healthy living can backfire. Dieting, for example, almost invariably fails to bring about the desired result of sustained weight loss. The sense of personal failure that comes as the scales lurch upwards again can produce a damaging drop in self-esteem and a sense of losing control; the frustrated dieter's response may be to abandon the diet and thus swing back to even greater porkiness. Mother Nature also conspires against the earnest dieter. People whose body weight oscillates because of dieting have a greater risk of premature death from coronary heart disease or other causes. Unsuccessful dieting can be bad for your health – and most dieting is ultimately unsuccessful.

What of behavioural self-destruction in literature? Fiction is littered with protagonists who recklessly expose themselves to danger, neglect their health or run themselves into an early grave because of great unhappiness or emotional turmoil.

An early case history of self-destruction appears in *Le Morte d'Arthur*, Sir Thomas Malory's fifteenth-century version of the legends of King Arthur and the knights of the Round Table. It is the sad tale of the Fair Maiden of Astolat and her doomed love for Sir Launcelot.

The brave, noble, irresistibly attractive Sir Launcelot rides to Astolat en route to a joust, and stays the night there at the home of the elderly baron, Sir Bernard of Astolat. Sir Bernard has a beautiful and virginal young daughter, the Fair Maiden of Astolat, who is at once smitten by Sir Launcelot. She is, as Malory so engagingly puts it, 'hot' in her love for the noble knight: 'for he is the man in the world that I first loved, and truly he shall be last that ever I shall love.' (Astolat, by the way, is Guildford and the maiden's name is Elaine. Fortunately, 'The Fair Maiden of Astolat' has more Arthurian resonance than 'Elaine of Guildford'.)

Sir Launcelot is grievously wounded and the Fair Maiden goes to look after him. Night and day she tends him, until his wounds are healed and Sir Launcelot is ready to take his leave. The Fair Maiden of Astolat beseeches Sir Launcelot to marry her or, failing that, at least go to bed with her. But the upstanding knight will not countenance marriage and refuses to dishonour the Fair Maiden by indulging in extramarital frolicking. She begs him again to be her husband or her lover, but to no avail. '"Alas," said she, "then must I die for your love."' The noble knight leaves Astolat to get back to some real man's work (fighting), leaving the emotionally wrecked Fair Maiden of Astolat behind him. Her mental state and self-destructive behaviour soon wreak havoc upon her physical health:

> Now speak we of the Fair Maiden of Astolat that made such sorrow day and night that she never slept, ate, nor drank . . . So when she had thus endured a ten days, that she feebled so that she must needs pass out of this world, then she shrived her clean, and received her Creator . . . 'it is the sufferance of God that I shall die for the love of so noble a knight . . . I loved this noble knight, Sir Launcelot, out of measure, and of myself, good Lord, I might not withstand the fervent love wherefore I have my death.'

True words from the Fair Maiden of Astolat, because very soon she dies. Clutched in her hand is a letter proclaiming her love for Sir Launcelot. That love has sent the Fair Maiden to her death, a death achieved through her behaviour.

Reckless behaviour allied with emotional distress can destroy an individual's physical health, as illustrated in *Jude the Obscure*, Thomas Hardy's novel about 'a deadly war waged between flesh and spirit'.

Jude Fawley, a self-educated young man of lowly origins, aspires to leave his unlovely country village and enter the hallowed portals of Christminster (Oxford) University. But the restrictions imposed upon Jude by class and poverty mean that he must instead make his way as a humble stonemason. Jude's romantic life is as frustrating and unsuccessful as his academic life. After being trapped into an ill-fated marriage to a pig-breeder's daughter he falls in love with his cousin Sue. The two are drawn together by an almost mystical affinity, but Sue leaves him to marry an older man. The two lovers are eventually united and live together, unmarried and condemned by society, in poverty and unhappiness. In the end Jude loses Sue, who returns to her husband.

Having failed to fulfil both his intellectual and romantic desires, Jude goes into physical and mental decline. Like many a nineteenth-century tragic hero, he succumbs to a consumptive illness which proves to be terminal. Jude's behaviour exacerbates his medical condition. With careless disregard for his health he makes a long journey on foot in the pouring rain to see Sue for the last time. She rejects his pleas and he returns to Christminster, physically and emotionally broken. But, as Jude explains to his former wife, he was fully aware of the risk to his health when he undertook the journey:

> I made up my mind that a man confined to his room by inflammation of the lungs, a fellow who had only two wishes left in the world, to see a particular woman, and then to die, could neatly accomplish those two wishes at one stroke by taking this journey in the rain. That I've done. I have seen her for the last time, and I've finished myself – put an end to a feverish life which ought never to have begun!

Eventually he dies, alone and neglected, not yet thirty years old. Hardy implicitly takes a multi-causal view of Jude's final illness, since environmental and constitutional factors play a role in it, together with psychological stress.[4] His emotional distress at losing Sue and at the death of their children acted as a trigger, but the illness also has antecedents in Jude's weak constitution and the harsh conditions he endured during his time as a stonemason:

> I was never really stout enough for the stone trade, particularly the fixing. Moving the blocks always used to strain me, and standing the trying draughts in buildings before the windows are in, always gave me colds, and I think that began the mischief inside.

Most of us die sooner than we have to because of the way we behave and the choices we make. Personally, though, I have some sympathy with Publilius Syrus, who two thousand years ago expressed the opinion that: 'They live ill who expect to live always.'

Mind over immune matter

> In this struggle Tarrou's robust shoulders and chest were not his greatest assets; rather, the blood which had oozed under Rieux's needle and, in this blood, that something more vital than the soul, which no human skill can bring to light.
>
> Albert Camus, *The Plague* (1947)

We turn now to a less visible, but no less important, mechanism by which the mind and body interact to affect health: the immune system. Among the most important developments in recent years has been the discovery of numerous biological pathways connecting the brain with the body's defence and regulatory mechanisms. Through these pathways the biological system that underlies our thoughts, emotions and behaviour – the brain – can exert a pervasive influence on the biological system that defends the body against most forms of disease – the immune system.

Our physical health depends critically on how well our immune system is functioning. One reason why a person suffering from psychological stress is more susceptible to colds and infections is because their immune system is less able to resist when they are exposed to disease-causing viruses or bacteria. In the following chapters we shall be exploring the manifold ways in which the mind and immune system affect each other. But before we do this we need to clarify a few basic issues.

So far I have referred rather sweepingly to the mind's effect on the immune system, as though the immune system were a homogeneous entity whose activities could be measured in a simple way, like temperature or blood pressure. In reality, the immune system is a breathtakingly complex and subtle entity whose intricate workings are still far from being fully understood. Immunology is one of the branches of science that has made the most spectacular leaps in understanding over the past thirty years, but it still has a very long way to go. To unravel how the mind influences physical health we must first establish what the immune system does and how it works.

UNDERSTANDING IMMUNITY

The immune system is one of the great wonders of nature, rivalled only by the brain in its intricacy and elegance of design. It is a multi-layered system of biological defences whose primary purpose is to defend the body from bacteria, viruses, fungi, parasites, toxins, cancerous cells and other disease-causing agents. The immune system is indeed a system, in the strict sense of the word: a highly complex and co-ordinated array of interrelated, interacting elements.

Like the economy of a nation, the immune system is not located exclusively in one place. In fact, the cells of the immune system are spread out all over the body. The majority are located in those organs whose purpose often seems slightly mysterious to the layperson: the thymus (located at the base of the neck); the spleen (below and behind the stomach); the lymph nodes (clumps of tissue in the armpit, groin, behind the ears and elsewhere); the bone marrow; the tonsils; and obscure backwaters of the gut (Peyer's patches and the appendix).

Immune cells are also to be found in the blood. These are the white blood cells (or leucocytes). Immune cells are carried in the bloodstream to locations in the body where they are needed, particularly sites of injury or infection. When an area of tissue is injured or infected an inflammatory response is triggered: the blood vessels swell up and become more permeable, thus increasing the supply of blood and immune cells to the damaged area.

There are numerous types of white blood cell, but here we are primarily concerned with the lymphocytes, which make up about a quarter of all white blood cells in humans. Lymphocytes can be subdivided into three main categories: B-lymphocytes, T-lymphocytes and natural killer cells.[5] The latter are capable of spontaneously killing certain virus-infected or cancerous cells.

The body is protected by layer upon layer of immune defences, rather like the proverbial onion. Simple accounts of the immune system (and this is a *very* simple account) usually divide its actions into two categories: the very clever and the mind-bogglingly clever; or, more conventionally, non-specific immune responses and specific immune responses.

Non-specific immune responses are the body's first line of defence against bacteria, parasites and other foreign material. Their basic purpose is to prevent potentially harmful foreign materials from entering the body in the first place, or to destroy them when they do enter. They achieve this without recognizing precisely which foreign material they are dealing with.

At the simplest level, non-specific defences include the physical barrier of the skin; the minute hairs called cilia in the respiratory tract and elsewhere which expel foreign particles from the body; and chemical defences such as stomach acid and bacteria-destroying enzymes in saliva and tears. A more sophisticated layer of non-specific immune defence is provided by various classes of white blood cells, notably the monocytes and neutrophils. These can ingest and destroy bacteria and foreign particles, a process known as phagocytosis (literally 'cell-eating'). They also help other white blood cells to kill micro-organisms, and produce vital chemical messenger molecules called cytokines which co-ordinate different aspects of the immune response.

Now we come to the mind-bogglingly clever part of the immune system: the part that can recognize and respond specifically to each and every type of foreign material it encounters. This is the specific (or acquired) immune response. It has the ability to make ultra-fine distinctions between material that forms part of your body and material of foreign origin – in other words, between 'self' and 'non-self'. This can be achieved because the immune system contains within it a detailed image of your body. Anything that deviates from this image, including some cancer cells, is recognized as foreign and attacked. A foreign substance that generates a specific immune reaction when it encounters the immune system is referred to as an antigen (short for *anti*body *gen*erator).

The ability to distinguish reliably between 'self' and 'non-self' allows the immune system to attack foreign material without harming your body's own healthy cells. Your immune system could detect the difference between a cell from your body and an apparently identical cell from my body. This is why transplanting tissue from one person to another can be such a tricky business, requiring the use of immune-suppressive drugs. In order to circumvent the immune response certain parasites have evolved the ploy of disguising themselves as the host's own tissue, tricking the host's immune system into regarding them as 'self' rather than 'non-self'.

It is helpful to subdivide the specific immune response into two main categories: humoral (or antibody-mediated) immunity; and cell-mediated immunity. Humoral immunity is concerned with attacking antigens that are floating around in the body fluids surrounding your cells as opposed to antigens inside the cells – hence 'humoral' from the old word 'humour', meaning bodily fluid. Humoral immunity essentially involves the production of antibodies by B-lymphocytes.

Antibodies are a type of protein molecule called the immunoglobulins.[6] They are the body's mainstay against bacterial infection. Each antibody is unique to one particular antigen, so there are as many types of antibody as there are antigens. When a B-lymphocyte meets the particular antigen to which it responds, it undergoes biochemical changes and starts producing multiple copies of itself, a process known as proliferation. The newly formed cells that result from this proliferation, called plasma cells, then secrete antibodies into the blood. These antibodies latch on to the antigen and, all being well, the antibody–antigen complex is then chomped up by a passing phagocyte.

Cell-mediated immunity, the other main variety of specific immune response, is primarily concerned with attacking antigens inside the cells – for example, viruses. It is also responsible for the body's immune reactions to transplanted tissues and tumours. Its main agents are the T-lymphocytes, which can recognize and kill target cells, such as those infected with viruses and foreign cells. Unlike antibodies, however, T-lymphocytes are unable to attack antigens that are floating around by themselves; instead they are dependent on other immune cells which 'present' the antigen to them, while at the same time stimulating the T-lymphocytes to attack by releasing chemical messenger substances known as cytokines. When stimulated in this way, T-lymphocytes proliferate and transform themselves into various subclasses with specific functions. Cytotoxic T-cells attack the antigen, while suppressor T-cells and helper T-cells regulate the whole delicate process. They do this by producing cytokines which alter the activity of other immune cells. Helper T-cells also stimulate B-lymphocytes to produce antibodies. The biological mechanisms regulating all of this are immensely complex.

The immune system learns and adapts each time it encounters a new antigen, setting a pattern for the way it will respond should it meet that antigen again. This is why you can be immunized against certain diseases, such as polio, typhoid, tetanus, rabies, diphtheria and chickenpox, and why there are diseases you catch only once in a lifetime. In this respect the immune system is like the brain: it detects and responds to specific stimuli in the outside world and then forms a long-lasting memory of those stimuli.

Vaccination exploits these immunological memory processes. A harmless fragment or heat-killed version of the bacteria or viruses is injected into the body. The antigens in the vaccine trigger the production of antibodies, but not the disease. The immune system is thus better prepared when it encounters the genuine item. Some micro-organisms are able to keep changing their biochemical appearance, which prevents the immune system from learning about them. The viruses responsible for the common cold and influenza are particularly good at this trick, which is why we do not develop permanent immunity to colds and 'flu.

Many things can impair the effectiveness of the immune system, including genetic defects, drugs and disease. There is also a general decline in immune function with old age. Sleep deprivation and poor nutrition both have marked effects on the immune system, too. Experiments on volunteers have ascertained that two or three days of sleep deprivation will produce significant reductions in various aspects of immune function. Even modest disturbances in sleep patterns can bring about measurable changes in the immune system. A study of healthy male volunteers found that depriving men of sleep for a few hours between 3 a.m. and 7 a.m. was enough to lower the immunological activity of their natural killer cells by more than a quarter. A good night's sleep returned it to normal.

What happens when the immune system goes wrong? Although it is vital to our existence, most people have only a vague understanding of what the immune system does and seldom give it a thought until it malfunctions. If it fails to recognize and destroy potentially harmful agents such as bacteria, viruses or cancer cells the result may be a serious disease. Those born with defects in their humoral immune responses suffer from recurrent, severe infections.

AIDS is a vivid example of what happens when the immune system is damaged. The human immunodeficiency virus (HIV) wreaks its havoc mainly by destroying the victim's helper/inducer T-lymphocytes. The eventual outcome is the almost invariably fatal condition known as Acquired Immune Deficiency Syndrome, or AIDS. One of the hallmarks of AIDS is a dramatic fall in the number and activity of helper/inducer (CD4) T-lymphocytes, though HIV does affect the immune system in other ways as well. An individual whose immune system is crippled by HIV becomes easy prey for a range of opportunistic infections and tumours, such as pneumonia, tuberculosis, Kaposi's sarcoma and non-Hodgkin's lymphoma, and it is usually one of these that kills the victim in the end.

AUTOIMMUNITY

In addition to destroying potentially harmful antigens, the immune system must be able to identify and avoid attacking its own body. Discriminating accurately between 'self' and 'non-self' is a fundamental requirement. There are times, however, when this discrimination fails for some reason and the immune system starts attacking 'self'. B-lymphocytes manufacture antibodies against other cells in the body and these autoantibodies start to attack healthy tissue. The result is an autoimmune disorder. Those who liken the immune system to an army repelling foreign invaders have used the illuminating metaphor of 'friendly fire' to describe the phenomenon of autoimmunity.

Autoimmunity is thought to play a role in at least twenty (and perhaps in excess of forty) diseases, and the list is growing. Among those included are rheumatoid arthritis; various thyroid disorders such as Graves's disease and Hashimoto's disease; primary biliary cirrhosis of the liver; systemic lupus erythematosus; Guillain-Barré syndrome; multiple sclerosis; diabetes mellitus; uveitis; pernicious anaemia; myasthenia gravis; and inflammatory bowel disorders such as celiac disease, ulcerative colitis and Crohn's disease.

The mechanisms of autoimmunity are not fully understood. Certain autoimmune diseases appear to arise because cells become altered in various ways – perhaps by viral infection or mutation – so that the immune system no longer recognizes them as 'self'. In other cases, autoimmunity results from a failure in the complex and delicately balanced mechanisms that regulate the immune system.

Genetic factors are known to play an important role in some autoimmune disorders. For example, insulin-dependent diabetes mellitus (otherwise known as childhood-onset diabetes) results from the autoimmune destruction of cells in the pancreas, the organ responsible for producing the blood sugar-regulating hormone insulin. Diabetics have a genetic predisposition to develop the disease. But environmental factors, including stress, also play their part in determining whether a genetically predisposed individual actually develops the disease. Even if one of a pair of genetically identical twins has diabetes there is still a 50–70 per cent chance that the other twin will *not* get the disease. Chronic psychological stress significantly increases the risk that those who are genetically predisposed will advance to full-blown diabetes.

Females are much more susceptible than males to a number of autoimmune diseases, and this is true both in humans and other species. Women are three times more likely than men to suffer rheumatoid arthritis, six times more likely to develop autoimmune thyroiditis, and at least ten times more likely to suffer from systemic lupus erythematosus. This sex difference in disease susceptibility is at least partly a consequence of hormonal differences; male and female sex hormones, such as testosterone, progesterone and oestradiol, influence the immune system in various ways.

Autoimmune diseases result from excessive or inappropriate immune activity. Accordingly, they may be ameliorated by drugs that suppress the immune system – the opposite of what happens in normal infectious diseases. This is why doctors use immune-suppressive drugs to treat autoimmune diseases.

MEASURING IMMUNITY

A prerequisite for any research into the relationships between mind, immunity and health is the ability to measure how well (or how badly) the immune system is doing its job. But as we know all too well, the immune system is not a simple entity whose activity can be readily described by a single index, any more than the multifaceted complexities of human intelligence can be fully encapsulated in a single number called IQ. So how do scientists measure immunity?

The functioning of a highly complex system such as a national economy can be quantified, albeit rather crudely, but only by using a diverse range of measures to describe various aspects of the system. Thus economists have come up with a variety of indices for quantifying economic activity, including the gross domestic product and gross national product; the headline and underlying rates of inflation; assorted interest rates; indices of money supply; the trade balance between imports and exports; exchange rates against foreign currencies; various measures of unemployment (most of them controversial); foreign debt; government borrowing; gold and currency reserves; measures of consumer spending, and so on. Each measure says something different about the economy and no single one gives a complete picture of the whole system.

Your immune system is far more complex than any national economy. Therefore if it is simplistic to talk of a national economy going up or down, it is even more simplistic to talk of an immune system going up or down. Fortunately immunologists, like economists, have at their disposal a number of informative ways of assessing certain basic aspects of this complex system.

Much of the research on how psychological and emotional factors affect immune function has focused on the white blood cells, primarily because they are the easiest to get at. Studying what is going on inside the thymus or spleen is difficult and intrusive, but taking a blood sample is quick and painless. Modern techniques also allow scientists to measure antibodies in saliva, one of the most readily accessible bodily fluids.

Scientists assess the immune function of white blood cells in two basic ways: by counting particular types of cells to see how many there are circulating in the bloodstream, or by measuring how well those cells perform their immunological functions.

The simplest approach is to count the total number of white blood cells in a given volume of blood, although this produces a crude index of limited use. An improvement on this is to count a specific type of white blood cell, such as T-lymphocytes or natural killer cells, or to gauge the amount of a particular class of antibody. But the biological and medical meaning of these measures is not always clear. A drop in the number of circulating lymphocytes may simply mean that lymphocytes have been shunted elsewhere in the body, probably into the spleen. At any one time only about 10 per cent of all lymphocytes are circulating in the bloodstream; the rest are stored in lymphoid organs such as the spleen and lymph nodes. Further information is provided by calculating the relative proportions of various cells, such as the ratio of helper T-cells to suppressor T-cells, since these proportions must be about right for the immune system to function properly.

The second, more revealing, approach relies on what are known as functional measures. These assess how well the various cells are performing their immunological functions. Research into the relationships between psychological factors and immunity has tended to rely on two particular functional measures of immune activity: the responsiveness of lymphocytes to stimulation and the cell-killing activity of natural killer cells. Because these two measures are so central to research in this field they warrant closer inspection.

As we have already seen, lymphocytes will respond to antigens by proliferating or producing multiple copies of themselves. This response can also be triggered in a non-specific way (that is, in the absence of the specific antigen) by chemicals known as mitogens. Lymphocyte function can thus be measured in a test tube simply by introducing mitogens to the sample.[7] Responsive lymphocytes will proliferate wildly when stimulated with the appropriate mitogen (a Good Thing), while unresponsive lymphocytes will be sluggish (a Bad Thing). Assume this is what I mean when henceforth I refer to lymphocyte responsiveness.

The prime function of natural killer cells is to destroy certain types of virus-infected cells and cancer cells. The obvious method of assessment in this case is to see how effective they are at destroying suitable target cells in the test tube. This is what is meant by natural killer cell activity, the second workhorse measure of immune function that will crop up repeatedly in subsequent chapters.[8]

It has to be admitted that lymphocyte responsiveness, natural killer cell activity and most of the other measures commonly used to assess immune function are flawed, allowing only a partial glimpse into the complexities of the immune system. Scientists can no more encapsulate the state of an individual's immune system in a few numbers than they can sum up that person's behaviour or emotional state with a few numbers.

It is a curious and regrettable fact that scientists tend to place greater faith in measurements of reassuringly physical entities, like the responsiveness of lymphocytes or the activity of natural killer cells, than they do in measurements of supposedly abstract entities like thoughts, emotions or behaviour. There is no rational basis for this prejudice, which is yet another reflection of our propensity to regard mind and body as two fundamentally different sorts of thing.

Despite their appearance of objectivity and precision, immunological measures are not totally reliable and, to a degree, depend upon subjective judgements. Conversely, the common prejudice that all psychological measures are by definition subjective, unreliable and woolly is a gross fallacy. Some are, but others are not. Experimental psychologists and behavioural biologists have been devising valid, reliable ways of measuring behaviour, psychological states and emotions for over half a century.[9] Psychological measures are not always perfect, but they are inherently no better or worse than immunological measures in terms of their reliability, validity and objectivity. (Here endeth the lesson.)

The mind–immunity connections

Scientists and doctors have traditionally tended to regard the immune system as an autonomous entity that operates to a large degree independently of the mind and behaviour; an exquisite piece of biological machinery honed by millions of years of evolution to act autonomously in protecting the body from anything the outside world can throw at it. But this view is now known to be fundamentally wrong.

The body's three main regulatory systems – the central nervous system (which includes the brain), the endocrine system (which produces hormones) and the immune system – do not work in isolation from one another. On the contrary, they are intimately connected and interact with each other in many important ways. Events occurring in the brain can produce changes within the endocrine and immune systems through a variety of routes, including specialized nerve pathways and chemical messengers. The effect may be to impair or enhance aspects of immune function, with potential consequences for health. The central nervous system, immune system and endocrine system are part of an integrated regulatory network that helps to ensure the survival and effective functioning of the whole organism. We are not just a bag of bits.

Another common misconception is that organisms work in a top-down, hierarchical manner with commands flowing in a single direction, from the brain to the rest of the body. In reality, information flows both ways along the various biological pathways that connect the central nervous system, endocrine system and immune system. Activity within the immune system can therefore influence the brain, mental state and behaviour.

Some scientists have likened the immune system to a sort of sensory organ, which is distributed throughout the body and provides the brain and endocrine system with information about the internal and external environment. This analogy makes good biological sense. The immune system detects the presence of antigens, including cells of its own body which have undergone change. It then transmits this information to the central nervous system, together with information about its immune response.

The new field of scientific research which is concerned with the complex inter-relationships between psychological and emotional factors, the brain, hormones, immunity and disease goes by the jaw-bending name of psychoneuroimmunology. In the next chapter we shall examine a few examples of how the brain and immune system interact. Before we do that, however, let us glance at the nature of the mechanisms which connect the brain and the immune system.

Since the 1980s, psychoneuroimmunologists have made considerable progress in understanding the biological pathways by which the brain and immune system influence each other. These pathways are of two basic sorts: electrical pathways using nerve connections, and chemical pathways using hormones, neuropeptides and other chemical messenger molecules. The specialized nature of these mechanisms strongly implies that they have evolved for a purpose – to enable the brain and the immune system to communicate with each other.

One good reason for believing that the brain and the immune system are meant to communicate is that they are hard-wired to each other by nerve connections. The tissues of the immune system are connected to the central nervous system by a rich supply of nerves. These nerve connections are responsible, amongst other things, for helping to regulate the development, activity and movement of lymphocytes and other immune cells.

Immune tissues in the spleen, bone marrow, thymus, lymph nodes, tonsils and gut are all abundantly supplied with nerve endings. Bone marrow, for example, is connected to the central nervous system by nerves emanating from the spinal nerve which supplies that region of the body. Some of the nerve connections in immune tissues do have other purposes, such as helping to regulate the local flow of blood, but some nerves undoubtedly pass information between the immune tissues and the brain. The spleen has a dense network of nerve connections and at least half of them are involved in transmitting information to and from the brain.

The second principal way in which the brain and immune system communicate is through an array of special chemical messenger molecules. The central nervous system and the immune system – the body's two great regulatory and memory systems – have a lot of chemical communications hardware (or should I say wetware) in common. That such vast numbers of these chemical messengers have been discovered indicates their importance as a mechanism of communication between the central nervous system and the immune system, and also within each system.

Neurotransmitters, hormones and other chemical messenger molecules that were once thought to be restricted to the brain and nervous system are now known to be active within the immune system as well. Conversely, certain immunotransmitters and other chemical messengers that were once thought to be exclusive to the immune system are now known to act on the endocrine and central nervous systems. The brain and the immune system speak the same languages.

Cells of the immune system have special biochemical receptor sites on their surfaces which respond specifically to chemical messengers produced by the central nervous system. Lymphocytes and other immune cells respond to a range of neuropeptides, neurotransmitters and hormones which are either produced directly by the central nervous system, or whose secretion is under its control. These chemical messengers include noradrenaline, corticosteroid hormones, endorphins, encephalins, growth hormone, adrenocorticotrophic hormone (ACTH), prolactin, substance P, substance K, vasoactive intestinal peptide (VIP), angiotensin and somatostatin. A number of these chemical messengers travel to the immune system via the blood circulation; others, like neuropeptides, are also delivered locally from nerve endings.

These chemical messengers are able to modulate aspects of cell-mediated and antibody-mediated immunity. For example, the hormone noradrenaline, which is released from the adrenal glands (under stimulation from the brain) and from nerve endings, has widespread effects on immune function. Noradrenaline can facilitate the production of antibodies in various immune tissues; it can also inhibit the division of lymphocytes and impede the destruction of virus-infected cells or cancer cells by the immune system. Another messenger molecule, substance P, makes lymphocytes more responsive to stimulation, increases the production of certain types of antibody and facilitates the movement of lymphocytes to sites of infection.

We shall see in chapter 5 how psychological stress stimulates the release of hormones, including cortisol, a steroid hormone which suppresses various aspects of immune activity. The stress-induced release of cortisol is controlled by a region of the brain called the hypothalamus. The hormone prolactin is also released in response to psychological stress; unlike cortisol, however, its main effect on immune activity is stimulative.

Chemical communication between the central nervous system and immune system also works both ways. The immune system sends chemical messages to the brain. Immune cells produce neuropeptides, hormones and other chemical messengers, including ACTH, endorphins, encephalins, VIP and growth hormone, which influence both the endocrine and central nervous systems.

Some of the most important messenger molecules mediating the communication between the central nervous system and immune system are the cytokines. These were originally thought to be exclusive to the immune system, but they are now known to act on the central nervous system and endocrine system as well. Scientists have found that cells in several regions of the brain and central nervous system either contain cytokines or have receptor sites for them. When cytokines are released by activated immune cells they can have widespread effects on an organism's nerves, hormone levels and psychological state. For example, when an infection occurs the cytokine interleukin-1 (IL-1) acts on the brain to induce slow-wave sleep and loss of appetite; IL-1, acting in concert with interleukin-6 (IL-6), induces fever by modulating the temperature control centres in the brain – in effect, putting the body's thermostat on a higher setting. IL-1 and IL-6 help to make you feel hot, sleepy and indifferent to hunger when you are ill. Cytokines are also active within the endocrine system. The cytokines IL-1, IL-2, IL-6, interferon-gamma and tumour necrosis factor (TNF) are all capable of influencing the release of hormones by the pituitary and adrenal glands.

Small changes to the structure of the brain will produce corresponding changes within the immune system, thus providing further evidence of communication between the two. Highly localized damage (or lesions) to parts of the brain, including the limbic forebrain, hypothalamus, brain stem and cerebral cortex, can bring about specific changes in immune function. Small lesions in the anterior hypothalamus produce reductions in lymphocyte responsiveness, natural killer cell activity and antibody production, while lesions in limbic forebrain structures such as the hippocampus and amygdala can increase the responsiveness of T-lymphocytes. Brain lesions can even modify certain immune-mediated diseases. For example, tiny lesions in the anterior hypothalamus alter the growth of tumours and allergic responses. In addition, animals with certain hereditary defects in their central nervous system are more vulnerable to immunologically-mediated disorders such as arthritis.

A final and fundamental point, to which we shall be returning in subsequent chapters, is that the electrical and chemical communication pathways between the central nervous system and the immune system operate in both directions. Information about the state of the immune system can be passed up to the brain and hence influence the organism's psychological state.

There is abundant experimental evidence that changes in immune activity are accompanied by corresponding changes in hormone levels, nerve activity and psychological state. Experiments have further revealed that patterns of electrical and chemical activity in the hypothalamus, limbic forebrain and other brain regions are linked to changes in immune activity which occur during the course of an immune response. For example, the peak production of antibodies in reaction to an immunological challenge (inoculation) is accompanied by changes in the electrical activity of nerve cells in the hypothalamus and other parts of the brain. Our brains appear to know, at least to some extent, what is going on within our immune systems.

In conclusion, then, we have seen that our psychological and emotional state can shape our perception of health and hence our sickness-related behaviour. At the extremes we may, like Colin Craven, feel ill and demand medical attention even though we have no disease; or, like Henry Earlforward, we may deny the reality of our symptoms and allow a serious disease to advance unchecked.

But our minds do far more than alter our perception of reality: they alter reality itself. The mind can affect our susceptibility to real physical diseases by modifying our behaviour or by directly influencing our immune defences, to which it is connected via electrical and chemical communications pathways.

By means of these psychological and biological mechanisms our minds really can make us ill.

4

Mind and Immunity

> O the mind, mind has mountains; cliffs of fall
> Frightful, sheer, no-man-fathomed. Hold them cheap
> May who ne'er hung there.
>
> Gerard Manley Hopkins, *No worst, there is None* (1885)

Evolution has equipped our bodies with psychological and biological mechanisms which enable the brain and immune system to talk to each other. Let us now examine what happens when these mechanisms operate in practice. In this chapter we shall look at what the immune system can do to the mind and, conversely, what the mind can do to the immune system. We shall see that changes in a person's mental state can affect their immune function and vice versa. Then we have the remarkable phenomenon of immune conditioning, which teaches the immune system to respond to purely psychological stimuli. We shall explore the curious connections between left-handedness, developmental learning disorders and immunological diseases. To round off the story we shall study in closer detail one specific instance of how mind, body and health interact, by considering herpes virus infections.

What can the mind do to the immune system?

Let us start with the most basic of psychoneuroimmunological questions. (The question is much simpler than its epithet.) What effects do psychological factors actually have on the immune system?

Probably the first ever scientific account of a psychoneuroimmunological phenomenon appeared in 1919, when a Japanese scientist called

Ishigami published the results of his research on tuberculosis in schoolchildren. Ishigami observed that an increase in tubercular illness coincided with a period when the children and their teachers were experiencing high levels of 'emotional excitement'. Using a crude immunological measure – the ability of white blood cells to destroy foreign bacteria – Ishigami was able to relate this upsurge in disease to a decline in immune function. He concluded that the emotional stress was responsible for the decline in immunity which, in turn, led to the increased incidence of disease.

Since Ishigami's day, and especially since the 1980s, scientists have accumulated a huge and varied pool of evidence that psychological and emotional factors influence the immune systems of humans and other species, with consequential effects on physical health.[1] It is time to look at a few examples.

BEREAVEMENT AND NUCLEAR DISASTERS

The death of a spouse or partner is one of the most devastating forms of psychological and emotional disturbance anyone can experience. It is also surprisingly common, affecting more than 800,000 people each year in the USA alone. By the age of sixty-five over 50 per cent of American women have been widowed at least once.

Bereavement considerably heightens the risks of death and disease for the surviving partner. The risks are particularly great if the surviving partner is male: widowed men in their twenties have a mortality rate seventeen times higher than comparable married men. Moreover, the increased mortality rates among widowers persist for several years after the death of their wife.

One factor in this increased disease and mortality among the bereaved may be that they undergo changes in their immune function. A number of surveys have found that the death of a spouse is followed by measurable reductions in the immune function of the surviving partner.

In a pioneering 1970s study of this phenomenon, scientists at the University of New South Wales in Australia took blood samples from recently bereaved people and measured how responsive their lymphocytes were to stimulation by mitogens. The results were compared against the immune function of non-bereaved individuals who were matched for age, sex and other relevant factors. Several weeks after the death of their spouse the bereaved subjects' T-lymphocytes were significantly less responsive.

Similar conclusions emerged from a prospective study which investigated the immune function of men whose wives had terminal breast cancer. During the weeks immediately following bereavement the widowers' lymphocytes were not as responsive as they had been beforehand. In the majority of cases their lymphocyte function eventually returned to normal, but some men were still exhibiting reduced immune function a year later. The changes in lymphocyte responsiveness could not simply be explained away as by-products of behavioural changes which might also have followed bereavement, such as disrupted sleep, inadequate diet or increased smoking.

Bereavement is also followed by a drop in natural killer cell activity. Scientists have found reductions in the natural killer cell activity of women whose husbands have died from lung cancer within the previous few months and in women whose husbands are in the final stages of lung cancer.

The death of a partner is obviously not the only way in which marriages or long-term relationships are disrupted. Separation and divorce also result in the severing of ties and can have an effect on the partners' immune function and health. Indeed, the health risks arising from separation and divorce are, if anything, worse than those associated with bereavement.

The stress which attends the loss of a partner is as old as the human race, but the twentieth century has ushered in new anxieties, such as the prolonged fear of exposure to dangerous radiation following a nuclear accident. Sadly, this is now a well-documented phenomenon. One of the most detailed scientific investigations of chronic stress has been conducted on people living near the damaged Three Mile Island nuclear power plant in Pennsylvania.

On 28 March 1979 the Three Mile Island reactor suffered a serious accident. Fortunately, the damage was contained and there was no significant release of radiation into the surrounding environment. However, the accident understandably generated great anxiety and distress among local inhabitants. In its immediate aftermath there was panic and confusion; saturation reporting by the media fanned the flames of fear that the reactor core would melt down and cause a massive release of radiation. Even after the initial crisis had passed, locals were painfully aware of the long-term clean-up operation and the continuing threat posed by the damaged reactor. Fear, anxiety and stress persisted for years.

Six years after the Three Mile Island incident American scientists measured the immune functions of people living within a five-mile radius of the damaged reactor. Compared with matched control subjects living further away, those living near Three Mile Island had markedly impaired immune systems. To be specific, they had fewer circulating B-lymphocytes, fewer natural killer cells, fewer suppressor/cytotoxic T-lymphocytes and a reduced immunological control over latent herpes viruses. And to round off the picture, they also showed a higher degree of anxiety and psychological distress; higher blood pressure; higher average pulse rates; and higher levels of the stress hormones adrenaline, noradrenaline and cortisol.

The Three Mile Island incident demonstrated quite clearly that, irrespective of whether or not there is a release of harmful radiation, the psychological stress which attends a nuclear accident can still have a major impact on mental and physical wellbeing. A further general conclusion from this and similar studies is that the immune system does not always adapt to long-term stress. Prolonged stress can produce a prolonged depression in immune function. It can also have some subtle effects on disease patterns.

The plot thickened. Three years after the original accident a team of researchers from Columbia University discovered an unusually high incidence of cancers among local inhabitants. In 1982 the incidence of all types of cancer among those living near the plant was 50 per cent above the historical norm. However, the cancer rate then started to fall, and by 1985 it had dropped *below* the long-term average. Meanwhile, cancer rates among the control subjects living further away from the plant had remained roughly constant. What was going on?

A number of fine minds were brought to bear on this problem. They concluded that subtle psychological processes lay at its root. All the evidence indicated that this peculiar blip in the cancer rates was not merely a consequence of exposure to radiation released from the damaged reactor. For a start, very little radiation had been released and the levels were too low to have caused a measurable increase in cancer. Moreover, there was no discernible association between the recorded levels of radioactivity and the pattern of cancer occurrences in the Three Mile Island area. The consensus among scientists was that the cluster of cancers around Three Mile Island could not be attributed directly to radiation.

Instead, it appeared that the transient increase in cancer rates probably stemmed from the high degree of psychological stress local residents were experiencing. Psychological surveys confirmed that the closer a subject lived to the plant, the higher on average their levels of anxiety and distress.

Psychological stress could have generated the apparent rise in cancer rates in two distinct ways. First, the local population had been bombarded with publicity about the potentially dire health consequences of the incident. They were more anxious about their health in general and particularly concerned about the dangers of radiation-induced cancer. It is therefore probable that they would have paid closer attention to their own health – especially to any possible symptoms of cancer – and would have sought medical attention at an earlier stage than individuals who were less concerned. This process would explain why cancer rates initially rose and then fell below the long-term average. The overall long-term incidence of cancer stayed roughly constant, but in the years immediately after the accident cancers were being diagnosed at an earlier stage in the disease – hence the blip.

Assuming this hypothesis is true, the Three Mile Island incident might actually have helped to save the lives of a few people who would have died had their cancer progressed undetected.

A second and more direct route by which the Three Mile Island accident could have influenced cancer rates is through stress-induced changes in immune function. The prolonged stress of worrying about the hazards posed by the damaged reactor might have led to an impairment in immune function, thus increasing locals' susceptibility to cancer or, more plausibly, promoting the growth of existing, early-stage tumours.

The Three Mile Island incident was to prove a pale shadow of what took place seven years later in the Ukraine. The Chernobyl nuclear disaster in April 1986 resulted in vast quantities of radioactive material being released into the environment. Thousands of square miles were contaminated and the inhabitants exposed to dangerously high levels of radiation which claimed hundreds of lives. Stress contributed to the problem, making an appalling situation worse. The official scientific report produced for the United Nations International Atomic Energy Agency commented that there were very high levels of psychological stress even outside the contaminated regions.

Another nuclear accident that caused serious psychological and physical damage, albeit on a smaller scale than Chernobyl, occurred in Brazil the following year. In September 1987 an unsuspecting thief stole twenty grams of caesium-137, a highly radioactive isotope, from a disused clinic in Goiania, a town in central Brazil. Before the true nature of the theft was discovered, the strange substance which glowed in the dark had been handed around the neighbourhood. Tragically, 118 people were contaminated with caesium-137, of whom 4 died from the immediate effects of radiation. And, as at Chernobyl, the entire population of the surrounding area was confronted with the dreadful possibility that they might have been exposed.

When the Brazilian authorities finally discovered what had happened they set about clearing the area of radioactive contamination. Entire houses and trees were removed, along with truck loads of contaminated topsoil. The resulting forty tons of low-level radioactive waste had to be put somewhere and the authorities chose to store it under armed guard in a town called Abadia. The waste storage site was within a mile of neighbouring homes.

In Abadia, as at Three Mile Island, the townsfolk had thus far suffered no actual exposure but suddenly found themselves facing the threat that they could be exposed if anything were to go wrong in the future. Goiania, on the other hand, *had* been exposed to radiation and feared its possible effects. Furthermore, there was little or nothing anyone could do about the situation.

Three and a half years after the Goiania incident, scientists investigated its impact on the communities of Goiania and Abadia. The results showed that the impact had been severe and long-lasting.

Residents of Goiania who had been exposed to radiation during the original incident were still suffering from the physical and psychological consequences of the trauma. Their blood pressure was raised; they had high levels of the hormones adrenaline and noradrenaline (indicating chronic stress); and they suffered from palpitations. They exhibited an abnormally high number of gastrointestinal disorders and other physical complaints. Their mental state was also disturbed: they reported continual feelings of fear and anxiety, found it difficult to concentrate, and their ability to make decisions had been impaired.

Not all of these symptoms could be accounted for by the physical effects of radiation. The psychological impact of the accident also appeared to play a substantial role.

This view was borne out by what the scientists found among Abadians living near the waste storage site. Although they had not been exposed to any radiation, they exhibited similar deteriorations in physical and mental health. If anything, they were in worse condition than the Goianians who had been irradiated. Much of the damage appeared to stem from the psychological impact of the accident rather than its physical effects.

SPACEFLIGHT, EXAMS AND OTHER NASTINESS

Danger and uncertainty can affect immune function even in highly trained and motivated individuals who have willingly exposed themselves to the hazard. Since the first days of the manned space programme it has been recognized that the briefest of spaceflights can produce immune suppression and an increased risk of infection. Astronauts exhibit various stress-related changes in the numbers and proportions of their white blood cells, including a marked decrease in the number of circulating lymphocytes and the responsiveness of those lymphocytes.

The immunological consequences of another exotically stressful situation were investigated by Norwegian scientists from the University of Bergen. They measured what happened to the immune systems of men who were learning how to evacuate an oil-drilling platform. The training procedure involved climbing a twenty-metre tower and being sealed into a lifeboat which was then tilted at 35 degrees before being allowed to free-fall into the sea, twenty metres below. The procedure was repeated on four consecutive days. The participants had never done this before and found the experience both uncomfortable and stressful.

Plummeting from a great height into the sea inside a sealed lifeboat evoked a number of changes in the men's hormone levels and immune function. Scientists measured an increase in two stress-related hormones (cortisol and prolactin) and in two immunological indices (the concentrations of immunoglobulin IgM and complement component C3 in their blood). But by the time of their fourth drop the men had grown accustomed to the experience and their biological response was accordingly less pronounced than it had been on the first three days.

Note that in this instance a brief trauma was accompanied by an increase in immune function, not a reduction. Emphasis has traditionally been placed on the immune-suppressive effects of stress. However, acute stress can actually enhance the activity of the immune system. Stress is not all bad. We shall be returning to this theme in the next chapter.

Psychological distress need not result from anything as severe as bereavement or as physically hazardous as spaceflight to impinge upon the immune system. Even a mundane activity like taking an academic examination can have an impact.

We all know that exams are a dismal experience and tales abound of students becoming ill as a result. Exams offer researchers a predictable, ethical and naturally occurring source of psychological stress, while medical students make co-operative and convenient subjects. Not surprisingly, a lot of research has been done on how exams affect the immune systems of medical students. The information wrung from the blood of these hapless students consistently reveals a drop in cell-mediated and antibody-mediated immune function at around exam time.

In an elegant series of psychoneuroimmunological studies, Janice Kiecolt-Glaser, Ronald Glaser and their colleagues at Ohio State University measured the immune function of medical students before, during and after taking their final exams. The measurements showed that examination-related stress was accompanied by numerous changes in immune function.

Both the number and the immunological activity of natural killer cells dropped significantly at exam time, in comparison with baseline levels measured one month before the exam. There were also marked reductions in the responsiveness of lymphocytes to stimulation by mitogens and viruses; the ability of T-lymphocytes to kill virus-infected cells; immunological control of three types of latent herpes viruses; and the primary antibody response to vaccination. The production by lymphocytes of the cytokine interferon-gamma fell by a factor of 25 at exam time. Furthermore, exam stress led to changes in the numbers and proportions of various categories of T-lymphocytes and in the protective response of white blood cells to irradiation with gamma rays. Changes also occurred at the genetic level: the expression of various genes in the students' white blood cells fell at exam time, including genes coding for the cytokine IL-2 (an important immune system messenger molecule). There was no evidence that the multifarious alterations in immune function were simply a reflection of changes in the students' eating or sleeping habits.

These alterations in immune function at exam time were paralleled by changes in how the students perceived their mental and physical health. Perhaps predictably, they felt distressed and reported suffering more illnesses (particularly minor respiratory infections) during the exam period than at other times in the academic year.

Cast your mind back to your own experience of taking exams and you will doubtless recall that stress and anxiety did not suddenly materialize on the day the exams began. You probably started feeling anxious at least a few days beforehand. Consistent with this, further research on exam stress has uncovered reductions in immune function in the build-up to the actual exam, seemingly related to the stressful anticipation. The lymphocyte responsiveness and antibody production of highly stressed psychiatry students showed a marked fall two weeks before a critical fellowship exam, but returned to normal within a few weeks after the exam.

Anticipatory responses like these illustrate an important general principle. Merely thinking about something unpleasant can itself be stressful. The unpleasantness in question can be in the past or in the future: recollection of an old trauma or apprehension about impending disaster. Laboratory experiments have confirmed that asking subjects to recall a stressful experience from their past is enough to provoke brief reductions in their immune function.

Human memory and conscious thought have the dubious benefit of greatly expanding the time-scale over which psychological and emotional factors can affect immunity. This ability may be vital in helping us to survive, plan ahead, learn from our mistakes and avoid hazards, but it can have its drawbacks. Human immune systems are more vulnerable to psychological and emotional influences than those of species whose mental states are rooted in the here and now.

Whilst exam-related stress is mercifully short-lived, other common forms of stress, such as being unemployed or trapped in an unhappy marriage, can persist for years. Short-term (acute) stress and prolonged (chronic) stress differ in their effects on immune function and health. To study the impact of chronic stress and the ability of victims to adjust to such continual pressure, Janice Kiecolt-Glaser and her colleagues investigated the immune function of caregivers looking after a spouse or relative with Alzheimer's disease. Alzheimer's is a debilitating progressive disease which can take anything up to twenty years to run its full course, during which time the sufferer will grow increasingly dependent – a state of affairs that places incessant physical and emotional demands on the caregiver for years on end. It is well established that those who care for relatives with Alzheimer's are themselves at greater risk of psychological and physical illness. But what does the stress of being a caregiver do to the immune system?

Kiecolt-Glaser's group found marked impairments in the immune systems of full-time caregivers. They had fewer circulating T-lymphocytes and poorer immune control over latent herpes viruses. Their natural killer cells were less responsive to stimulation by cytokines. They also experienced a higher incidence of physical illness (especially respiratory infections), psychological distress and depression. Revealingly, the caregivers who had the least social support and who were most upset by their spouse's dementia displayed the biggest reductions in immune function.

Connections such as these between psychological stress and altered immune function are certainly not unique to humans; they have been extensively investigated in other species. Many forms of stressful situation, including overcrowding, handling by humans, loud noise, or physical restraint, have been found to alter the cell-mediated and antibody-mediated immune responses of animals. For instance, mice or rats subjected to repeated periods of physical restraint show a marked depression in their inflammatory and cell-mediated immune responses to viral infection, together with a delay in their antibody response. Stressed mice and rats have less responsive T-lymphocytes, reduced natural killer cell activity, smaller antibody responses to novel antigens and lower levels of cytokines such as IL-2 and interferon-gamma.

Returning to humans, one of the more surprising conclusions from research is that even commonplace situations which bear no comparison with the ordeal of, say, bereavement can have a measurable impact on immune function. Laboratory experiments with volunteers have confirmed that something as banal as performing a difficult mental task lasting less than half an hour can result in appreciable reductions in lymphocyte responsiveness and alterations in the numbers of circulating white blood cells. The links between mind and immunity can, it seems, intrude into everyday life.

There is growing evidence from psychoneuroimmunological research that the minor stresses and hassles of daily life can exert a variety of adverse effects on immune function in otherwise healthy individuals. One study, for example, found that day-to-day fluctuations in the amount of immunoglobulin IgA in the subjects' saliva correlated with fluctuations in their mood; IgA levels were higher on days when the subjects were in a good mood and lower on days when their mood drooped. Researchers at the University of Pennsylvania found that immune function was affected by normal levels of anxiety in healthy people who were not regarded as having any psychological or medical problems. Among a sample of healthy, averagely happy and seemingly unstressed student volunteers, individuals with more anxious dispositions achieved lower scores on two indices of immune function (lymphocyte responsiveness and the circulating level of IL-1). It all starts to seem slightly alarming – or does it?

DOES IT MATTER?

The evidence for connections between psychological factors and immune function seems overwhelming; the examples described above are only the tip of an iceberg. But before we leap too far ahead we ought to consider a fundamental question that is often posed by sophisticated sceptics. Psychologically-induced changes in immune function may well exist – but do they have any tangible impact on our health? In medical parlance, are the effects of psychological factors on immune function *clinically* significant (as opposed to statistically significant)?

There is no doubt that a gross impairment of immunity leads inexorably to disease: AIDS is a vivid example. But the reductions in immune function that are associated with psychological and emotional factors are often modest in size, and it is not at all obvious that they will have much impact on physical health. A drop in immunological indices is not an infallible sign that illness is on the way.

Just as it would be simplistic to describe a virus as the sole and sufficient cause of a disease, so it is equally simplistic to think of changes in immune function in that way. Diseases have multiple causes. Two individuals can experience identical fluctuations in immune function, yet what tips the balance in favour of developing a disease in one case may have no ill effects in the other. Psychological influences on immune function often have a greater impact on the health of the elderly or infirm than they do on healthy, robust youngsters.

We need to bear these caveats in mind at all times, in order to avoid making over-enthusiastic leaps of faith. Yet the conclusion that mind, immunity and health are interrelated stands, for there is abundant evidence to support it. Many of the immunological indices that are affected by psychological factors play an important role in disease. For example, a reduction in natural killer cell activity is associated with adverse consequences in several disease processes, including viral infections, autoimmune disorders and the progression of several cancers. Persistently low levels of natural killer cell activity result in a rise in infections. Similarly, a reduction in lymphocyte responsiveness is linked to increased health problems and mortality.

Still more conclusive, though, is the growing inventory of psychoneuroimmunological studies which have tied together all three elements in the story. These have demonstrated that psychological factors have an effect on immune function which in turn affects our physical health. We shall be considering this evidence in greater detail later.

What can the immune system do to the mind?

> O, what a noble mind is here o'erthrown!
>
> William Shakespeare, *Hamlet* (1601)

In the previous chapter we saw that the electrical and chemical communication pathways which connect the central nervous system to the immune system allow information to flow in both directions. The mind–immunity link is a two-way street.

The obvious implication is that changes within a person's immune system might influence their psychological state and perhaps their behaviour, too. Is there any evidence that this happens? Does immunity have a demonstrable effect on the mind? So far, research has barely scratched at the surface of this issue and the evidence is sparse. The focus has been predominantly on how the mind alters immunity rather than the other way round. Nonetheless, some light has been cast on the problem.

Simply being ill alters our psychological state and behaviour. Minor infections such as colds and 'flu are attended by measurable changes in memory and other aspects of mental performance, while severe infections are typically attended by lethargy, depression, loss of appetite and general malaise. Usually these psychological and behavioural changes assist the process of recovery and keep the sick individual out of harm's way. Far from being incidental by-products of disease, they are mediated by specific biological mechanisms. The immune response to infection triggers the release of chemical messengers (cytokines) which act upon the central nervous system to induce malaise, drowsiness, fatigue, an increase in slow-wave sleep and loss of appetite. These changes accompany illness by design, not accident.

A number of diseases resulting from immunological disorders have specific psychological as well as physical consequences. Over 50 per cent of those who suffer from the autoimmune disease systemic lupus erythematosus (SLE) develop distinctive psychological deficits, such as memory loss, impaired mental performance and emotional disorders. Investigations of SLE-prone mice suggest that the immunological changes associated with the disease are also responsible for the psychological changes.

Viral infections can have a more direct impact on behaviour, especially if viruses get into the central nervous system. For example, a mild infection with herpes simplex virus during the first few days of life can have long-lasting effects on the behaviour and learning ability of mice. These psychological changes are thought to arise because viral infection alters the development of the brain.

AIDS is a more familiar example. Although AIDS is generally thought of as quintessentially a disorder of the immune system, the HIV retrovirus can find its way into the central nervous system and cause damage there as well. The end result, now recognized as a characteristic symptom of AIDS, is a form of dementia in which the victim suffers impairments to thought processes and muscular control and, in extreme cases, severe psychiatric symptoms such as paranoid psychosis.

Aside from causing physical damage to the brain, HIV infection has other, less brutally direct ways of affecting the mind. The knowledge, or even the mere suspicion, that they are infected with HIV can generate enormous emotional distress in people. High levels of anxiety and depression are common among HIV-positive individuals and in high-risk groups.

Over the years there has been recurrent interest in the hypothesis that certain mental illnesses, including schizophrenia, might be caused by a virus. This hypothesis is obviously appealing, offering as it does the tantalizing prospect of magic bullets to cure complex mental diseases. So far the evidence for viral causation has been at best inconclusive. Links have, however, been established between psychotic mental illness and peculiarities in immune function.

Studies of schizophrenics and those with a family history of schizophrenia have uncovered assorted abnormalities in subjects' cell-mediated and antibody-mediated immune function. Of particular note is the discovery that some schizophrenics produce antibodies which attack their own brain tissue (autoantibodies). This discovery led to a theory – as yet unconfirmed – that schizophrenia might be an autoimmune disease, caused by the immune system attacking the brain. More recently, there has been renewed interest in the idea that autoimmune disorders could lie at the roots of multiple sclerosis and Alzheimer's disease. But for the time being the jury is still out.

The causal connections between mental illness and immunological abnormalities remain open to doubt. There are plenty of less remarkable ways in which mental and immunological abnormalities might come to be linked. For example, infection with herpes simplex virus (HSV) is much more prevalent among patients in mental hospitals than in the general population, but this does not prove that HSV is a cause of mental illness. A simpler explanation is that the unusual behaviour and institutional lifestyle of mental patients increase their chances of being infected with HSV and other viruses. The picture is further complicated because patients suffering from serious psychotic illnesses are usually on powerful drugs, which might themselves be responsible for any immunological abnormalities.

DEPRESSION

Another issue that remains tantalizingly uncertain is the relationship between depression, immune function and physical health.

Ill people are often depressed. This is hardly surprising. Being ill is a depressing business, especially if your illness is debilitating or life-threatening. Depressed people also have a higher mortality rate than those who are not depressed. Again, this need not indicate anything terribly profound; for a start, depressed people are statistically more likely to have accidents or commit suicide. But perhaps there is something more interesting going on here as well. Could depression make us more susceptible to physical illness? Sick people get depressed, but do depressed people get sick?

Regardless of any immunological consequences, depression is a formidable public health problem in its own right. At any one time, around 2–3 per cent of the population is suffering from a serious depressive illness. If depression also increases our vulnerability to infections and other diseases then clearly it becomes an even greater problem.

There is substantial evidence that individuals suffering from severe depression (as opposed to merely feeling a bit glum) have a higher probability of falling ill or dying prematurely than non-depressives. A twenty-year study of two thousand middle-aged American men revealed that those who showed signs of depression had twice the risk of developing a fatal cancer in later years, irrespective of other medical risk factors such as smoking or a family history of cancer.[2]

One way in which depression could increase susceptibility to illness would be by impairing the body's immune defences. If psychological stress can have this effect, why not depression? The available evidence indicates that there is indeed a connection between depression and impaired immune function, although the relationship is not a simple one.[3]

Researchers have uncovered an array of immunological changes associated with depression. Among the most consistent of these are reductions in natural killer cell activity, lymphocyte responsiveness and immune control over latent herpes viruses. Severe depression is also accompanied by changes in the numbers and proportions of the various white blood cells, and an increase in the production of cytokines such as IL-1 and IL-6. Severe melancholic depressives have fewer T- and B-lymphocytes, an increased ratio of helper to suppressor/cytotoxic T-lymphocytes and more monocytes.

Depression is also associated with changes in hormone levels akin to those found in stress (of which more in chapter 5). In particular, depressives tend to have higher levels of the stress hormones cortisol, adrenaline and noradrenaline, all of which have an impact on the immune system. These hormonal changes probably play a mediating role in the relationship between depression and altered immune function.

The picture that emerges is a complex one. Several of the changes in immune function that accompany depression are characteristic of immune suppression, but many elements of the immune system are *more* active in depressives. Indeed, depression has been likened to a sort of inflammatory disorder.

Another route which would allow depression to influence immunity is by changing behaviour patterns. On average, depressed people get less sleep, take less exercise, eat less healthy food, smoke more cigarettes, drink more alcohol and use more drugs (both therapeutic and recreational). As we saw in chapter 3, all of these behaviour patterns affect immunity and health in their own right. A few studies have indicated that the reduction in natural killer cell activity which often accompanies clinical depression might stem from disturbances in sleep patterns and physical activity. Unscrambling the precise mechanisms whereby depression acts upon immune function is an immensely difficult problem.

What of the imaginary world of literature? The seemingly impenetrable web of interconnections between mental and physical illness is strikingly evoked in *Crime and Punishment*, Fyodor Dostoyevsky's tale of wretchedness, crime and sickness in nineteenth-century St Petersburg.

The impoverished former student Raskolnikov murders an old money-lender and her sister. But why? Does Raskolnikov commit the murder in the name of his nihilistic philosophy, to affirm his personal superiority? Does he do it simply because he is destitute and needs the money? Or is it the irrational act of a sick man whose mind is deranged by a physical illness? *Crime and Punishment* has lent itself to countless literary interpretations, so why not a psychoneuroimmunological interpretation?

In the period leading up to the murder Raskolnikov contemplates and plans his crime. He has spent weeks alone in his claustrophobic rented room, isolated from society, crushed by poverty and suffering from a profound malaise. He barely eats or drinks, sleeps only fitfully and is in a state of almost constant anxiety:

> for some time now he had been in a tense, irritable state of mind that verged upon hypochondria . . . It would have been hard to go much further to seed or to sink to a lower level of personal neglect.

He develops a fever. His hands tremble and his mood oscillates between restlessness and physical exhaustion:

> He was pale, his eyes were burning, he was suffering from exhaustion in every limb . . . an extraordinary, feverish and somehow helpless turmoil seized hold of him, in place of his slumber and torpor.

Raskolnikov is obviously a sick man before he commits the crime. Whether his sickness is of the mind or body is unclear, but the picture is consistent with that of a man suffering from a serious physical illness which has upset the balance of his mind.

Then comes the crime. After Raskolnikov has murdered the old woman (and, unexpectedly, her sister as well) his condition goes from bad to worse. As he contemplates the horror of what he has done Raskolnikov degenerates into sickness and delirium:

> Now he was suddenly attacked by an ague so violent that his teeth nearly leapt from his mouth, so violently did they chatter, and his entire body started to shake . . . 'I think you're quite ill, aren't you?' Nastasya observed . . .

As Raskolnikov lies gripped by delirium, lassitude and violent mood swings, his worried friends gather around him. One of them, a doctor, concludes that the origins of Raskolnikov's illness lie deeper than the appalling physical conditions in which he has been living for the past few months:

> It is, so to speak, the product of many complex mental and material influences – of anxieties, apprehensions, worries, of certain ideas . . .

Raskolnikov soon falls under police suspicion, not least because he publicly draws attention to the possibility that he might be the murderer. He enters into a protracted game of psychological cat-and-mouse with the state investigator, Porfiry Petrovich. They discuss at length Raskolnikov's mental and physical state. Though Raskolnikov maintains that he has always acted on the basis of conscious, rational decisions, Porfiry is not convinced:

> It's an illness, Rodion Romanovich, an illness! You're neglecting your health, sir. You ought to consult an experienced medic . . . What's wrong with you is delirium! The plain fact is that you're doing everything in a state of delirium!

Later, as he contemplates his impending doom, Raskolnikov himself begins to consider this possibility:

> It's these really stupid, purely physical ailments, which are linked to the sunset or something, that make a man do stupid things, no matter how hard he tries not to!

On the verge of mental and physical collapse, Raskolnikov hands himself over to the law and confesses to the murders. At his trial the court refuses to believe that he committed the crime solely for financial gain or in the name of some abstract philosophical theory, despite Raskolnikov's own assertion that he did it for the money. The court hears of the accused's delirium and 'inveterate hypochondria', and his sick and impoverished condition prior to the crime. Everyone is convinced that Raskolnikov could only have murdered the women under the influence of temporary insanity. So, instead of being executed Raskolnikov receives the relatively lenient sentence of penal servitude in Siberia.[4]

Immune conditioning

A second strand of evidence to support the interaction of the mind and the immune system comes from the remarkable phenomenon of immune conditioning. What this shows is that the immune system can learn, by association, to respond to psychological stimuli such as a novel taste or smell. The individual's immune system can learn to respond to, say, the sweetness of saccharin, the smell of camphor, or the manifold stimuli associated with the laboratory, in much the same way as it would respond to a powerful immune-suppressive drug.

Immune conditioning works in essentially the same manner as behavioural conditioning – the process by which Pavlov's famous dogs learned to salivate at the sound of a bell which had previously signalled food – and it is perhaps best described by analogy with Pavlov's experiments.

In Pavlov's archetypical experiment dogs were repeatedly exposed to two stimuli: the first of these (food) automatically evokes a behavioural or physiological response (salivation), the second (the sound of a bell) is biologically neutral and initially evokes no such response. By consistently ringing a bell to signal the arrival of food, Pavlov conditioned the hungry dogs to associate the bell with food. Soon the sound of the bell by itself was enough to elicit the same response – salivation and eager anticipation – as the food.[5]

By exact analogy, immune conditioning involves pairing a biologically neutral stimulus which initially has no effect on the immune system (such as an unusual taste or smell) with a stimulus which does affect the immune system (such as an immune-suppressive drug). Once conditioning has taken place, the taste or smell can elicit the change in immune response without the drug having to be administered. It is as though the immune system has learned that this particular taste or smell signals a change in immune function and responds accordingly.

Astonishingly, the phenomenon of immune conditioning was first unearthed by Russian scientists in the 1920s, when Pavlov's influence was at its peak. Guinea pigs were injected with an antigen, thereby provoking an inflammatory immune response. This immunological stimulus was repeatedly paired with a biologically neutral stimulus such as scratching or heating a patch of skin. Once a guinea pig had been conditioned in this way, the heat or touch stimulus alone was enough to elicit the immune response.

This pioneering work on immune conditioning was duly forgotten about until the 1950s, when it was temporarily revived in the Soviet Union only to be forgotten again, as sometimes happens in science when an empirical discovery does not fit into an established theoretical framework. At that time there was no satisfactory way of explaining how it was possible for immune responses to be conditioned and many scientists simply dismissed the data.

Half a century later two American scientists – a psychologist called Robert Ader and an immunologist called Nicholas Cohen – serendipitously rediscovered immune conditioning. Their resurrection of this phenomenon was a landmark in the development of the new field of research that became known as psychoneuroimmunology.

Ader and Cohen were studying a special form of behavioural conditioning known as learned taste aversion.[6] As part of their research Ader and Cohen gave rats a novel-tasting drink (saccharin-sweetened water) and then injected the rats with a drug called cyclophosphamide to make them nauseous. The rats soon learned, through Pavlovian conditioning, to avoid sweet-tasting water because it made them sick. But then rats which had been conditioned in this way began to die in unexpectedly large numbers, and it seemed that the more sweet water they had drunk the more likely they were to die.

This was puzzling. How could a biologically neutral stimulus like sweetness make a rat die? Ader and Cohen knew that besides making rats feel sick, cyclophosphamide also suppresses the immune system. In fact, it is one of the drugs administered following tissue transplants to prevent immune reactions. They therefore hypothesized that a form of conditioning process might have occurred, in which the previously neutral stimulus (the sweet taste) had come to elicit the same biological response as the drug, namely immune suppression.

Ader and Cohen duly put their hypothesis to the test. They conditioned a number of rats by giving them saccharin-sweetened water paired with injections of cyclophosphamide. Three days later they gave the rats more sweet water, this time on its own. To test the rats' immune responses, Ader and Cohen injected them with red blood cells taken from a sheep. Foreign blood cells are powerful antigens and provoke the production of antibodies.

The results were clear cut. Rats that had been subjected to immune conditioning produced significantly fewer antibodies. The sweet taste by itself was sufficient to reduce their antibody response by a quarter.[7] Since Ader and Cohen's ground-breaking experiment in the 1970s, the finding that changes in immune function can be conditioned and subsequently evoked by psychological stimuli has been replicated time and again, with many variations on the basic theme.

Psychoneuroimmunologists have demonstrated that immune conditioning can alter numerous aspects of cell-mediated and antibody-mediated immunity, including lymphocyte responsiveness, natural killer cell activity, cytokine production, antibody levels, white blood cell numbers, histamine production and the immune response to foreign tissue grafts. A wide assortment of psychological stimuli besides saccharin-sweetened water have been conditioned; for example, the smell of camphor, the taste of vanilla and even the complex of stimuli associated with the laboratory environment itself. Immune function has been manipulated by various means, including antigens, sundry forms of stress, and a whole range of drugs. For instance, sounds or visual cues that were previously paired with stressful, immune-suppressive electric shocks have been used to elicit a conditioned reduction in immune function. It has even proved possible to condition an association between a novel-tasting soft drink and the allergic immune reaction to pollen and house mites.

A string of mundane, alternative explanations for the observed results have been tested experimentally and eliminated. Immune conditioning is not simply a beguiling glitch in the data: it is a robust, reproducible phenomenon that works under a wide variety of circumstances and in different species.

The majority of such experiments have involved suppressing rather than enhancing the immune system, not least because it is relatively easy to suppress immune function with drugs. Conditioning an increase in immune function is an altogether trickier proposition. Nonetheless it has been demonstrated in certain forms, including the conditioned enhancement of antibody production and natural killer cell activity. Scientists have been experimenting with a substance called Poly I:C which enhances natural killer cell activity by stimulating interferon production. By pairing Poly I:C with an odour, it has proved possible to invoke a conditioned increase in natural killer cell activity in response to the odour stimulus alone.

More recently, conditioned immune enhancement has been demonstrated in humans. In one experiment, German researchers injected healthy volunteers with adrenaline, which evokes a transient increase in natural killer cell activity. Adrenaline injections were paired with a neutral, psychological stimulus: a combination of white noise and a sweet taste (sherbet). As predicted, the noise/sherbet stimulus was subsequently capable of evoking the increase in natural killer cell activity by itself.

IMMUNE CONDITIONING AND DISEASE

Immune conditioning can do more than modify immune function. It can also produce measurable changes in health. Experiments have demonstrated that immune conditioning can alter the progress of arthritis, lupus, cancer and an expanding assortment of other diseases.

In the early 1980s Ader and Cohen investigated what immune conditioning could do to systemic lupus erythematosus (SLE), a serious autoimmune disease which affects the skin, connective tissue and internal organs. You may recall from chapter 3 that autoimmune diseases arise when, to put it crudely, the immune system gets out of hand and starts attacking the body's own tissues. Accordingly, autoimmune diseases are often treated with drugs that suppress the immune system.

Ader and Cohen therefore investigated whether conditioned immune suppression could slow down the progression of SLE in mice. They conditioned the mice by pairing a neutral stimulus (saccharin-sweetened water) with weekly injections of an immune-suppressive drug (cyclophosphamide). Sweet water was then substituted for half the drug injections. The results showed that, in conditioned animals, the sweet water had much the same effect as cyclophosphamide. It significantly retarded the development of the disease.

In subsequent experiments scientists have demonstrated that conditioned immune suppression can slow the progression of arthritis (another autoimmune disease) and delay the rejection of skin grafts or transplanted tissue. Predictably, conditioned immune suppression has the opposite effect on 'normal' diseases in which the immune system plays a defensive rather than offensive role; for example, it can accelerate the growth of certain malignant tumours.

Immune conditioning effects would be of enormous value in the treatment of disease if they could be exploited clinically. One possible benefit would be a reduction in the doses of toxic immune-suppressive drugs needed to treat patients suffering from autoimmune diseases or after transplant operations. There is already at least one successful case history, in which conditioned immune suppression enabled doctors to reduce the doses of immune-suppressive drug given to a child suffering from lupus erythematosus. More exciting yet – though so far elusive – is the prospect of using conditioned immune enhancement to boost the body's natural defences against disease.

Immune conditioning can occur inadvertently, outside the laboratory, when patients receive immune-suppressive drugs. This can be a real problem for cancer patients undergoing chemotherapy. The powerful cell-killing drugs used for treating cancer often have nauseating side-effects. Patients who repeatedly experience bouts of severe nausea associated with chemotherapy can become behaviourally conditioned and start to feel sick in anticipation of their next treatment. Psychological stimuli that have previously been associated with the chemotherapy, such as the sight, sound, smell, or the mere thought, of the hospital make them feel physically sick. This anticipatory nausea occurs in anything from a quarter to three-quarters of chemotherapy patients and is believed to result from Pavlovian conditioning. As one physician wryly observed, 'I'd had more than a few patients myself who vomited at the sight of me and who had the decency to blame it on the chemotherapy.'

Chemotherapy can also produce conditioned changes in immune function. Research has revealed that cancer patients who receive immune-suppressive drugs can involuntarily learn to associate the hospital environment with the immune-suppressive chemotherapy. Once this immune conditioning has occurred, the hospital environment itself can elicit some of the immune changes that were previously produced by the chemotherapy.

One study of women receiving chemotherapy for ovarian cancer found a pronounced decrease in lymphocyte responsiveness immediately prior to the chemotherapy being administered. There were strong indications that conditioned immune suppression was to blame. A recent Swedish study found similar evidence for conditioned immune suppression in women undergoing chemotherapy for breast cancer, especially in individuals with higher levels of anxiety. Prior to chemotherapy there were reductions in the women's natural killer cell activity, the number of circulating monocytes and the ratio of their helper to inducer T-lymphocytes.

Once again the evidence lends support to the hypothesis that under certain circumstances patients' immune systems can become suppressed by the mere sight, sound, smell or thought of the hospital.

SOME ALLERGIC HISTORY REVISITED

Before we leave the subject of immune conditioning we shall revisit a selection of historical curios, armed with the knowledge of modern science. The discovery of immune conditioning has cast new light on some curious medical tales of yore, in which arbitrary psychological stimuli were apparently able to trigger an immune reaction.

Perhaps the best-known case is that of the so-called rose cold, accounts of which date back hundreds of years. A sixteenth-century Dominican monk is said to have been 'seized with syncope' (fainting) whenever he saw roses, no matter how distant. In the nineteenth century a doctor described how an artificial rose could provoke asthmatic symptoms in a patient who was allergic to roses, thereby demonstrating the role of 'purely psychic impressions' in the allergic reaction.

Assuming these and other similar accounts are based on factual observations, they probably reflect a conditioned immune response. Purely psychological stimuli such as the sight of a rose – even an artificial one – could elicit a conditioned immune reaction in allergic individuals.

Similar stories abound in the medical literature. In 1930, for example, one doctor described the case of a patient with a hay fever allergy so severe that an attack could be triggered merely by looking at a picture of a hay field. A further case history from the 1950s records how one patient exhibited allergic symptoms when shown a picture of a goldfish bowl. Laboratory experiments carried out during the 1950s confirmed that purely symbolic (that is, non-allergenic) stimuli such as visual images could reliably elicit asthmatic or allergic reactions in individuals for whom the stimuli were associated with genuine allergens. The same effects were achieved with animals, proving that the phenomenon could not be attributed to subjects cheating or allowing over-active imaginations to get the better of them. In the light of current knowledge we can be reasonably confident that these reactions were a product of immune conditioning.

Research has demonstrated that the immune response to the Mantoux or tuberculin skin test can also be modified by conditioning. This test is used to determine whether a person has ever suffered from, or been exposed to, tuberculosis. A small amount of tuberculin – a protein extract obtained from tubercle bacilli – is injected beneath the skin, usually on the forearm. If that person has been exposed to tuberculosis in the past a patch of inflammation will appear around the site of the injection within two or three days, indicating a degree of immunity to the tuberculin.[8]

In one conditioning experiment, tuberculin was administered to healthy volunteers on six consecutive occasions at monthly intervals. On each of the first five occasions tuberculin was taken from a green vial and injected into one arm, and an inert liquid (saline solution) was taken from a red vial and injected into the other arm. Over the course of the five conditioning trials the subjects unconsciously came to associate a positive inflammatory reaction with the green vial (tuberculin) and a zero reaction with the red vial (saline).

Then, without the knowledge of either the subjects or the nurse who was administering the injections, the researchers switched the contents of the two vials. For the sixth and final trial each subject received tuberculin from the red vial, which had previously been associated with a zero reaction. This time, the reaction to the tuberculin injection was significantly reduced, suggesting that some form of conditioned immune suppression had occurred. An unconscious expectation that an injection from the red vial would produce no effect was sufficient to reduce the inflammatory reaction to real tuberculin. Comparable effects have been demonstrated in mice, showing that conscious thought is not essential.

The strange story of the left-handed brain

We move now to a third strand of evidence concerning the myriad links between mind and immunity – the discovery that left-handed people are statistically more likely to suffer both from psychological problems and disorders of the immune system.

In the early 1980s two neurologists, Norman Geschwind of Harvard University and Peter Behan of Glasgow University, published some intriguing findings. They had discovered an unusually high incidence of immunological disorders and developmental learning disorders (such as dyslexia) among strongly left-handed individuals and their immediate relatives.

Geschwind and Behan compared a large sample of strongly left-handed people with a large sample of strongly right-handed people. The left-handers had more than double the incidence of immune disorders, especially bowel and thyroid complaints, and ten times as many developmental learning disorders such as dyslexia and stuttering. The first- and second-degree relatives of left-handers also had higher rates of immunological and learning disorders. Furthermore, Geschwind and Behan found a higher proportion of left-handers among hospital patients suffering from immune disorders or severe migraine.

The results of Geschwind and Behan's work initially seemed bizarre, but they did tie in with fragmentary observations from earlier research. For example, it had previously been noticed that people with autism or stuttering are more prone to allergies and bowel disorders; and that males and left-handers account for an unusually high proportion of children with developmental learning disabilities such as dyslexia, stuttering, autism, hyperactivity and Tourette's syndrome.

What has all this to do with the connections between mind, immunity and disease? The answer lies in the theory that Geschwind and Behan put forward to account for this strange cluster of connections between immunological, neurological and behavioural characteristics. They hypothesized that correlations between left-handedness, immune disorders and developmental learning disorders stem from interactions between the brain, the immune system and various hormones that take place early in life, inside the mother's uterus.

To understand this theory you must be aware of an important fact about the human brain. The two hemispheres of the cerebral cortex (the outer part of the brain) differ in their capacity to learn and perform certain functions. This asymmetry is known as cerebral dominance. In most individuals the left hemisphere is dominant for language functions while the right hemisphere dominates for spatial functions. Because of a crossover in the nerve connections between the brain and the rest of the body, motor control in right-handed people resides in the left brain hemisphere and vice versa. In other words, right-handers are left-brained and left-handers are right-brained. Damage to the left hemisphere of the brain in early development can result in dominance being transferred to the right hemisphere, producing a left-handed person.

According to the Geschwind–Behan hypothesis, when foetuses are exposed to abnormally high levels of the male sex hormone testosterone in the uterus it preferentially delays the growth of the left cerebral hemisphere, resulting in left-handedness and language-related learning disorders such as dyslexia and stuttering. Testosterone also affects the development of the thymus and other immune organs, thereby increasing the risk of immune disorders later in life. The testes of male foetuses add to the level of testosterone, which could explain why left-handedness, developmental learning disorders and immune disorders are all more common in males than in females.

This theory, which has been considerably refined over the years, has generated a fair amount of controversy and is not universally accepted.[9] Not all the studies that followed have been able to replicate the original findings. Since Geschwind and Behan published their results, however, a wide range of evidence has accumulated in support of the general idea that there are links between left-handedness, immune disorders (including hay fever, eczema, asthma and hives), and developmental learning disorders. For instance, a study of Norwegian children uncovered firm associations between left-handedness, dyslexia and immune disorders. In another experiment, left-handers exhibited a greater susceptibility to common allergy-provoking substances such as dust and animal fur. It even turns out that left-handed schizophrenic patients are six times as likely to test positive for abnormal autoantibodies as right-handed schizophrenics.

A statistical analysis, which drew together data from over fifty thousand cases and controls, confirmed that left-handers are significantly more likely to suffer from allergies and asthma, run twice the risk of developing inflammatory bowel diseases such as Crohn's disease, and have a slightly higher overall incidence of autoimmune disorders than right-handers.

More controversially, it has been suggested that left-handed people die younger than right-handers. Stanley Coren at the University of British Columbia in Vancouver has drawn together statistical evidence indicating that left-handers have a shorter average life span than right-handers, and that being left-handed is associated with an increased risk of serious accidents, immune disorders, alcoholism, addiction to smoking, mental illness and birth complications.

Reassuringly (for me at least, because I am left-handed) Coren's claims about the perils of left-handedness are controversial, and other researchers have reported conflicting data. For example, psychologists at Durham University tested the putative connection between handedness and longevity by analysing the lifespans of six thousand British first-class cricketers who were born between 1840 and 1960. The cricketing archives, which systematically record whether a bowler was left-handed or right-handed, revealed that left-handed cricketers lived as long on average as right-handers. The cricketing data did, however, support Coren's claim that left-handers are more accident-prone. Although their average life-span was no shorter, a significantly greater number of left-handed cricketers died from unnatural causes. They were especially likely to perish in warfare.

Some even stranger statistical connections with left-handedness have been unearthed. Males born during the spring or early summer have a greater probability of being left-handed than those with birthdays in autumn or winter. The chances of being left-handed are also higher if you are a twin or if your birth was traumatic. Research at Harvard Medical School has revealed a slightly higher than normal incidence of left-handedness among homosexuals. Alcoholics and problem drinkers also include a disproportionate number of left-handers. There is even evidence linking left-handedness, dyslexia and immune problems with exceptionally high levels of musical ability.

Like all good biological phenomena, the links between handedness, behaviour and immune disorders are not unique to humans. Strains of mice that have been selectively bred to show a high degree of 'paw-edness' (that is, a strong behavioural preference for using one paw rather than the other) also have deficits in immune function, including unusually low levels of T-lymphocyte responsiveness and natural killer cell activity.

The ideas we have encountered in this chapter could be illustrated with numerous examples of disease processes which are affected by psychological state, via its impact on immune function. The example I have chosen, mainly because it is so mundane – both you and I probably have it – is infection with herpes viruses.

The wonderful world of herpes

Unlike marriage, herpes is for life
Anonymous

It is Sunday. Tomorrow you have that all-important meeting. You need to look your best. But as you peer bleary-eyed into the mirror you sense an ominous tingle in your lip. With a depressing thud you realize that by tomorrow morning, just when you want to radiate self-confidence, your mouth will be graced by a glistening outcrop of cold sores which will make you feel as attractive as a tapeworm. You know from experience that the unsightly blisters on your lip will turn into equally unsightly scabs and that at least a week, possibly two, will pass before you return to your unblemished state. Why you? Why now?

One school of thought has it that cold sores are caused by stress. Is this true? You have a vague feeling that they do tend to crop up at the worst times, particularly when you are under pressure at work or having a hard time at home. But if cold sores are all in the mind then what is responsible for those loathsome blebs that are rapidly forming on your face?

Cold sores (or herpes labialis, to use the medical name) were well known to the ancients and were described by the Greek physician Hippocrates nearly two and a half thousand years ago. Greek and Roman etiquette demanded that those who were suffering from an outbreak should desist from kissing.

The vast majority of cold sores are caused by a virus called herpes simplex virus type 1 (HSV-1). And a virus, to quote Peter Medawar's memorable description, is 'simply a piece of bad news wrapped up in protein'. Lurking inside the protein coat of this particular virus are approximately seventy genes, in the form of DNA strands, whose sole purpose in life is to make as many copies of themselves as possible.

HSV-1 has a close relative called herpes simplex virus type 2 (HSV-2) which usually attacks below the waist and is responsible for genital herpes. (By the way, whatever happened to genital herpes, that subject of inescapable media brouhaha in the early 1980s? The recorded incidence of genital herpes has continued to climb, but it seems to have been eclipsed in the public mind by that much greater scourge, AIDS.) The two viruses are so closely related that their territories are not entirely exclusive, and cross-infection can result from oral sex with someone who has a cold sore. HSV-1 is also responsible for an exotic skin infection called herpes gladiatorum which afflicts wrestlers.

Cold sores are hardly life-threatening and rank fairly low in the pantheon of diseases. Herpes viruses can, however, prove hazardous to AIDS victims or others whose immune systems are not working properly. The lesions and ulcers produced by HSV-1 will be larger, more painful and longer-lasting in such cases; there is also a risk that they will spread to the tongue, palate and other areas where cold sores do not normally occur. In newborn babies and people with suppressed immune systems, HSV-1 can occasionally spread to the central nervous system and cause severe problems such as viral encephalitis or meningitis.

Genital herpes, caused by HSV-2, is considerably more distressing than facial cold sores and is capable of causing complications if it erupts during the final stages of pregnancy. Genital herpes infections also increase vulnerability to infection with HIV, possibly because the herpes blisters offer an easy route of entry for the HIV retroviruses.

There are psychological consequences, too. Recurrent genital herpes can generate emotional distress and anxiety, disrupt sufferers' social relationships and sex lives, undermine their self-esteem and, in severe cases, cause clinical depression. The psychological damage is sometimes more disabling than the physical symptoms and may require treatment in its own right, in the form of counselling, relaxation therapy or stress management training. Here we have an example of an immunological problem (the failure of immune control over latent viruses) generating psychological problems, rather than the other way around. We are back to our two-way street.

HSV-1 and HSV-2 belong to a family of herpes viruses which are responsible for a range of moderately unpleasant illnesses in humans. The other members of the family are varicella zoster virus, which causes chickenpox and shingles; Epstein-Barr virus (EBV), which causes glandular fever (alias infectious mononucleosis); cytomegalovirus; human herpes virus 6 (HHV-6); and human herpes virus 7 (HHV-7). Cytomegalovirus is present in the majority of adults and normally produces mild, cold-like symptoms (though it can have serious effects in AIDS patients and others with compromised immune systems). HHV-6 is also widely distributed and has been identified as the cause of roseola, a common feverish illness of young children. HHV-7, which was first isolated in 1990, is believed to infect at least 85 per cent of the American population, apparently without causing any noticeable disease.

In case it is any consolation, you are in good company if you suffer from occasional cold sores. The majority of adults are infected with HSV-1. Estimates of its prevalence vary widely according to the population sampled and the method used to detect the virus, but a conservative estimate is that at least half of us harbour the virus. Some studies have found that by the time people reach their thirties more than 90 per cent have been infected with HSV-1, although a figure of 60–70 per cent may be more typical in modern industrialized nations. The risk of infection with HSV-2 depends to a degree on sexual activity. The more you do it the greater your chances of being infected (although, like AIDS, once can be enough). Around 80 per cent of prostitutes are reckoned to be infected with HSV-2, compared with less than 3 per cent of nuns. The number of cases of genital herpes has risen rapidly; in the USA, for example, it increased nine-fold between 1966 and 1981.

Although most adults are infected with HSV-1 or HSV-2 (or both), relatively few suffer from recurrent cold sores or genital lesions. Less than half those infected with HSV-1 ever experience an actual outbreak of cold sores. Among those infected with HSV-2, as few as a quarter develop genital lesions. Herpes simplex virus infection is a prime example of a condition where the disease-causing agent (in this case a virus) can be present in a person without their developing any disease. It clearly illustrates why the simplistic notion of 'one germ–one disease' is so inadequate.

To understand how it is possible to be infected but not fall ill, it is first necessary to understand a little about the biology of herpes viruses. Infection with HSV-1 usually occurs during childhood or early adulthood. The virus is transmitted by direct contact with body fluids and usually enters the body via the mucous membranes – the lips, mouth or, less commonly, the rectum or conjunctiva. The initial infection may be accompanied by mild 'flu-like symptoms and inflammation of the gums and mucous lining of the mouth, but it often goes unnoticed.

The herpes viruses live in the sensory nerves that supply the site of infection. After the viruses have entered the body they migrate up the long, thin sensory nerve fibres and pass into a ganglion (or sensory nerve bundle) where they set up home permanently, remaining inside the nerve cells in an inactive, latent form. HSV-1 usually lives in the trigeminal ganglion, which is situated behind the ear, while HSV-2 favours the sacral ganglion near the base of the spine.

From time to time – perhaps after an interval of many years, perhaps never – the viruses are reactivated. Like spawning salmon, they migrate back down the sensory nerve fibres and start reproducing inside cells near the surface of the skin. The usual result is an outbreak of cold sores or genital herpes.

The first sign that a cold sore is brewing is a tingling, throbbing, burning or itching sensation at the site of infection. Within twelve to twenty-four hours a small raised spot appears. This soon develops into a collection of tiny, fluid-filled blisters which contain a large quantity of viruses, white blood cells and the remains of dead cells. After a further six to twelve hours the blisters break open and weep. A hard crust or scab then forms before the site eventually heals. The whole process normally takes ten days or so. Cold sores can break out anywhere on the skin or mucous membranes, but generally appear on or near the lips.

At present the only effective treatment for herpes simplex virus infection is the anti-viral drug acyclovir. This works by blocking the biochemical processes which enable herpes viruses to replicate inside your cells. Acyclovir inhibits the synthesis of viral DNA. Applying acyclovir as soon as you feel the first tell-tale tingling might just prevent the cold sores or genital herpes from fully forming. But neither acyclovir nor any other drug that is currently available will eliminate the viruses from your body.

Many different things can trigger a recurrence of cold sores: ultraviolet light in bright sunshine, illness, menstruation, a minor injury to the nerve where the viruses live, or even an epidural anaesthetic. However, many recurrences have no obvious physical trigger and seem to be linked to psychological catalysts such as stress. Surveys confirm that most sufferers detect a relationship between feeling stressed, tired or run down and the recurrence of cold sores. But is there any scientific evidence for this?

Three American studies in the 1970s examined the relationship between stress and cold sores among student nurses. The data revealed that individuals who reported being chronically unhappy or experiencing unpleasant moods had significantly more outbreaks of cold sores over the following year. Conversely, those who had more satisfactory social lives and greater social competence suffered fewer outbreaks. Chronic unhappiness and lack of social support appeared to assist the reactivation of latent herpes viruses.

Psychological factors can also have a bearing on outbreaks of genital herpes. Over fifty scientific studies, dating back to the 1920s, have found consistent associations between psychological stress or anxiety and the subsequent recurrence of genital herpes. Individuals who experience frequent recurrences of genital herpes tend to have a personality profile which makes them particularly prone to stress and anxiety. The recurrence of genital herpes is also associated with loneliness (although one could be forgiven for speculating which is the chicken and which is the egg).

The precise biological mechanisms which allow herpes viruses to remain dormant for long periods and then suddenly reactivate are not fully understood. Even when HSV-1 is in its latent state at least one of the viral genes is still actively expressing itself inside the host's cells. What is known is that the immune system (and the T-lymphocytes and macrophages in particular) plays a critical role in ensuring that the viruses stay dormant. Anything that reduces the competence of the immune system may permit the latent viruses to start replicating.

There is now abundant evidence that psychological stress can impair the immune system's ability to control latent herpes viruses. One study, for example, found that there was a drop in suppressor/cytotoxic T-lymphocyte levels a week or so before subjects had a recurrence of genital herpes; this drop in immune function was typically linked to an episode of stress or depressed mood.

The reactivation and renewed production of herpes viruses provokes a response from the host's immune system, which starts to produce antibodies against the viruses. An increase in the level of herpes virus antibodies therefore indicates that the host's immune system has faltered in its efforts to subdue the virus. People whose immune systems are suppressed by chemotherapy typically have high levels of antibodies against herpes virus. This is why psychoneuroimmunologists sometimes use the level of herpes virus antibodies as an inverse measure of immune function (the more antibodies the poorer the immune function).

Numerous psychoneuroimmunological studies have found that diverse forms of psychological stress are accompanied by raised levels of herpes virus antibodies, indicating a poorer immune control over the latent viruses. Stressful situations associated with raised antibody levels include taking an important exam, being lonely, caring for an Alzheimer's sufferer, living near the damaged Three Mile Island nuclear plant, and getting separated or divorced. We considered several of these examples earlier in this chapter. Increased herpes antibody levels are also common in cases of clinical depression. For some sufferers, just the fear of another attack of genital herpes can be enough to provoke one.

One remarkable experiment in the 1940s demonstrated that cold sores could be induced by hypnosis. A patient suffering from a psychiatric disorder was told under deep hypnosis that she was run down and debilitated. Bingo – within twenty-four hours the unfortunate woman was festooned with multiple herpetic blisters. Some party trick.

In another experiment an experienced meditator demonstrated that she was able deliberately to modify her immune reaction to the herpes virus varicella zoster. Yet another study (albeit based on only one subject) managed to relate both the disease symptoms of HSV-1 infection and the underlying immunological changes to psychological stress; over a period of several months, episodes of psychological stress were accompanied by a reduction in immune function, followed by a recurrence of cold sores. At times of maximum stress the numbers of helper T-lymphocytes fell and cold sores subsequently erupted on the unfortunate subject's face. Stress can trigger the reactivation of latent herpes viruses in other species as well.

There is little doubt that the mind can – and does – alter immunity. Psychological and emotional factors in various guises can influence both cell-mediated and antibody-mediated immune function. The mind–immunity connections have been demonstrated and replicated in so many ways and in so many laboratories that they can no longer be considered even mildly controversial.

5

The Demon Stress

> The mind is its own place, and in itself
> Can make a Heav'n of Hell, a Hell of Heav'n
>
> John Milton, *Paradise Lost* (1667)

> In the world ye shall have tribulation
>
> St John 16: 33

The mental state that we most commonly associate with poor health is stress. To judge from the amount of media coverage, the demon stress seems to have become the latter-day scourge of Western life. Stress is a perennial topic of conversation and something of a status symbol among thrusting business executives. It even merits a brief mention in most modern textbooks of medicine, despite the almost inevitable absence of any reference to psychoneuroimmunology. The media and self-help industry offer unlimited (though sometimes conflicting) advice on how to cope with it. But before we can tackle the effects of stress we must first understand the beast. In this chapter we shall consider what stress is, how it works and what it does to us.

What is stress?

> There is nothing either good or bad, but thinking makes it so.
>
> William Shakespeare, *Hamlet* (1601)

To begin at the beginning: what do we mean by stress? This is no dry semantic quibble. Stress is a muddled concept and over-use has rendered it virtually meaningless.

In colloquial speech 'stress' can be used interchangeably to refer to two quite different things. It may describe an unpleasant or potentially harmful external force which places us under pressure. Media reports tend to imply that anything unattractive or irksome may be deemed 'stressful' and therefore potentially harmful, thus providing us with a legitimate excuse – if not a binding duty – to avoid it. Hence the current vogue for fretting about stress.

On the other hand, 'stress' can also describe our response to unpleasant or potentially harmful events – in other words, something going on inside the body. We feel anxious or depressed, therefore we are stressed. Walter Cannon and Hans Selyé, the scientists who carried out much of the pioneering research on the biology of stress in the first half of the twentieth century, regarded the stress response as the organism's way of maintaining its internal steady state in the face of external disturbances.

Fortunately, it is possible to resolve this muddle. Nowadays when scientists refer to psychological stress they usually mean something which encompasses both the external disturbances and the individual's internal response to those disturbances. In this sense, stress refers to the outcome of an interaction between an organism and the demands placed on it by the environment.

To put it more formally, we can define psychological stress as the state arising when the individual perceives that the demands placed on them exceed (or threaten to exceed) their capacity to cope, and therefore threaten their wellbeing. Although the brain plays a central role in the whole process, conscious thought need not be involved. Your brain can recognize stress without your having to think about it. So can a rat's or a monkey's brain.

To make this distinction clearer, the unpleasant or potentially harmful things happening in the environment are referred to as stressors, while the psychological and biological reactions they elicit are referred to as the stress response.[1] Stressors can be physical as well as psychological: injury, disease, surgery, undernourishment and extremes of heat or cold are all potent stressors. Healthy individuals who undergo even minor surgery, for example, show a measurable drop in their immune function. Our focus, however, will be on psychological rather than physical stressors, since these are the ones that affect most of the people most of the time in modern industrialized nations.

Tucked inside our definition of stress is a crucial concept. Stress depends on how we appraise both the demands placed on us and our capacity to cope with them. Psychological stress is as much a function of how we see the world as how the world really is. Perceptions are all-important – a notion that is memorably portrayed in this exchange between Hamlet and Rosencrantz:

> 'Denmark's a prison.'
> 'Then is the world one.'
> 'A goodly one, in which there are many confines, wards, and dungeons; Denmark being one o' th' worst.'
> 'We think not so my lord.'
> 'Why then 'tis none to you, for there is nothing either good or bad but thinking makes it so. To me it is a prison.'

Because it is a product of how an organism interacts with its environment, stress itself is neither a characteristic of the organism nor a characteristic of the environment. Individuals differ considerably in their responses to identical stressors, depending on factors that are unique to them, including their past experience, beliefs, education, personality, physical health, genetic make-up and social environment. As we shall see in the next chapter, for example, people who have rich networks of supportive social relationships are generally less vulnerable to stressors than those who are lonely and socially isolated.

A situation that is highly stressful for one person may be of little or no consequence – or even positively invigorating – for someone else. Stress, like beauty, is partly in the mind of the beholder.

Someone who is about to make their first parachute jump will probably exhibit a greater stress response than a seasoned parachutist (assuming they have so far managed to land without breaking both legs). An individual's previous experience of a hazardous situation greatly influences how their mind and body will respond to that situation the next time around.

The power of experience to transform perception is illustrated in Stephen Crane's *The Red Badge of Courage*. Set during the American

Civil War, it tells the story of a raw young Union recruit and his baptism of fire.

Henry Fleming is eager for glory and despite his mother's discouragement he enlists in the Union army. In the run-up to his first battle Henry contemplates the possibility that he might be too scared to fight, but rejects it. All his life he has dreamt of battle and now is his chance to relish it. After a number of false alarms Henry finally experiences what it means to encounter real, live enemy troops. All around him men are being killed or wounded. It is not how he had dreamt it would be. Just as he is congratulating himself that the battle is over, the enemy attack again. It is all too much for Henry Fleming. He throws down his gun and shamelessly runs like a rabbit.

> On his face was all the horror of those things which he imagined . . . Since he had turned his back upon the fight his fears had been wondrously magnified.

In his terror and confusion, Henry blunders about in a forest at the edge of the battlefield. He tries to question a fellow Union deserter, but the frightened man lashes out with his rifle, knocking him unconscious. On recovering his senses, Henry rejoins his regiment at the front. His comrades mistakenly assume that his head wound – his 'red badge of courage' – has been inflicted by an enemy bullet. His spirits are bolstered by their attitude towards him.

Before long, Henry is plunged back into the heat of battle. This time, however, he is a changed man. He has encountered the horror of war and survived. The experience has transformed him. Now, instead of fleeing in panic he fights like a wild cat. When the colour sergeant is shot, Henry grabs the regimental flag and leads a charge. He is commended for his bravery, and as he leaves the battlefield for the second time Henry contemplates the profound changes that have come over him:

> He had been to touch the great death, and found that, after all, it was but the great death . . . So it came to pass that as he trudged from the place of blood and wrath his soul changed. He came from hot plowshares to prospects of clover tranquilly, and it was as if hot plowshares were not. Scars

> faded as flowers . . . He had rid himself of the red sickness of battle. The sultry nightmare was in the past.

Scientific literature contains numerous illustrations of how the same stressor can affect different individuals in different ways. One American study looked at the hormonal effects of stress on medical students taking an exam (yes, them again). The exam stressor provoked significant hormonal changes, but only in candidates who perceived the exam as highly stressful. Those who felt less traumatized by the exam showed little change in their hormone levels. The biological response depended on the student's psychological perception of the stressor.

Another American study investigated the associations between stressful life events, psychological distress and immune function among healthy volunteers. Again, similar life events affected individuals in different ways; some were better able to cope than others. 'Poor copers' exhibited signs of distress, depression or anxiety in response to life-event stressors and had lower natural killer cell activity than 'good copers' who managed to endure comparable stressors without experiencing psychological ill-effects.

A common but fallacious assumption is that stress can be measured objectively in terms of our external circumstances. It has been said that the current obsession with stress is merely the self-obsessed whingeing of a pampered generation which has in fact been exposed to less stress than any previous generation in history. How can this generation complain of stress, so the argument goes, when they contemplate the horrific suffering of those who lived through wars or famines? Surely a bullying boss or a rocky marriage cannot begin to compare with being shot at, starved, bombed or held in a prisoner-of-war camp?

Yet assessing the stressfulness of a situation is not as straightforward as it may seem. People can become more upset, both psychologically and physiologically, about apparently trivial things than they do about serious problems. The extent to which we are disturbed by a stressor depends on several things, including our sense of personal control over circumstances and the support we receive from those around us. There is no simple relationship between our physical environment and the amount of psychological stress we experience.

For instance, research on American soldiers during the Vietnam War found that the level of stress hormones in certain servicemen was lower during active combat than when the men were off duty. This was especially true among highly trained, élite units with a strong sense of group cohesion. Personal control and social support make a big difference, as we shall see later. It is common wisdom in military medicine that the closer troops are to the 'sharp end' the less likely they are to complain of illness. (Violent injury and death are another matter, of course.) The harassed accounts clerk really can have a worse time of it than the marine commando.

A widely overlooked feature of stress is that it is infectious. As well as suffering from stress ourselves we can inflict it upon others through our behaviour and attitudes. In chapter 7 we shall see that when people are subjected to the long-term stress of unemployment or job insecurity their families also suffer.

Stress, then, is not merely something that happens to us, a force of which we are the passive victims. It is a product of how we appraise and respond to our environment. We are – or can be – active agents in the process. The practical implication is that by changing the ways in which we view the world, rise to challenges, or assess our own ability to cope, we can alter our susceptibility to stress.

The obvious way of mitigating stress is to eliminate the stressor at its source. If threatened by an escaped lion or a mugger, run away; if worried about failing an exam, work hard enough to be confident of passing. Unfortunately, many of our routine strategies for coping with stress do little to eliminate the source of the problem and can be positively harmful into the bargain. I am thinking here of those age-old stress-management techniques like smoking, drinking mind-numbing quantities of alcohol, popping pills or scoffing comfort foods.

Our definition of stress suggests an alternative approach to coping, which is to change our appraisal of the stressor. At the crudest level we can do this by denying the very existence of the problem. But denial can be dangerous, as we saw in chapter 3: refusing to face up to the possibility that you might have a serious disease could prove fatal if it prevents you from seeking prompt medical attention.

A less extreme solution is to shift your focus of attention away from irksome matters. This seems to be a popular strategy on crowded commuter trains. Squeezed in among the sweating bodies and unable to change the ghastliness of their physical environment, the passengers mentally distance themselves from it by daydreaming, anaesthetizing their minds with a personal stereo, or even reading a book. This type of psychological defence mechanism – for that is what it is – can blunt the unpleasant emotions and anxiety associated with a stressful situation, even though it clearly does nothing to change the reality.

A more active version of this strategy is sometimes referred to by psychologists as re-framing. It entails deliberately changing one's conscious perception of a situation by focusing on its positive, manageable aspects and identifying practical solutions or, failing that, coming to terms with the inevitable. Stress-management advisers employ the concept of re-framing when they teach clients to view their stressors in a new, positive light. Setbacks are magically transformed into stimulating new challenges and disasters become valuable opportunities for personal growth. Despite the management school aura, re-framing does have its uses. So too do the old nuggets of folk wisdom, such as setting realistic objectives, not being a perfectionist, keeping things in perspective and having a sense of humour.

The characteristic way in which we as individuals tend to view the world – our explanatory style – has a substantial bearing on our susceptibility to the slings and arrows of everyday life. There is compelling evidence that those of us whose world view is essentially pessimistic – regarding our problems as pervasive, long-lasting, insoluble and our fault – suffer worse damage from stress than those irritating optimists who always look on the bright side of life.

A series of remarkable long-term studies in the USA, which tracked subjects for over thirty years, found that pessimists were more prone to illness and died younger on average than optimists. These variations in health emerged years later, suggesting that the explanatory style preceded the poor health and not vice versa; it is unlikely that the pessimists became pessimistic because they were already unhealthy. Pessimists were also poorer achievers as students, both on the sports field and at work.

A pessimistic explanatory style is also associated with lower levels of immune function among the elderly. One study discovered that individuals who attributed the negative events in their lives to fixed, pervasive personality traits had lower ratios of helper to suppressor T-lymphocytes. Conversely, there is evidence that people with a generally optimistic outlook, positive expectations and high self-esteem have fewer illnesses. Optimists also recover from surgery sooner than pessimists and enjoy a higher quality of life thereafter.

Fiction has dealt extensively with the concepts of re-framing and explanatory style. An extreme case of optimistic explanatory style and re-framing is embodied in the eleven-year-old heroine of Eleanor H. Porter's children's classic *Pollyanna*.

The eponymous orphan Pollyanna Whittier has a secret strategy for coping with the countless sorrows and misfortunes that befall her and those around her. She deploys the Special Game that she learned from her father. This involves finding something to be glad about in every circumstance, no matter how dire and depressing it might seem, and displaying 'an overwhelming, unquenchable gladness for everything that has happened or is going to happen.'

By dint of using this simple re-framing strategy Pollyanna manages to remain astonishingly buoyant in the face of dreadful adversity. Before long, the lonely, sad and ill inhabitants of Beldingsville fall under the spell of Pollyanna's Special Game and their lives are transformed. Thanks to Pollyanna, the gloom-laden Mrs Snow, infirm for fifteen years, realizes that 'she had been too busy wishing things were different to find much time enjoying things as they were.' The saccharine child even melts the heart of her emotionally crippled maiden aunt. The local doctor is quick to recognize how much Pollyanna and her Special Game have benefited the mental and physical health of Beldingsville's inhabitants. He tells the nurse:

> I wish I could prescribe her – and buy her – as I would a box of pills; though if there gets to be many of her in the world, you and I might as well go to ribbon-selling and ditch-digging for all the money we'd get out of nursing and doctoring.

As a final episode in Pollyanna's catalogue of relentless misfortunes she is run over by a motor car. (Her stressful life-events score would be worryingly high by now, even for an optimist.) Pollyanna is paralysed from the waist down and an eminent specialist pronounces that she will never walk again. However, the Special Game enables Pollyanna to overcome her serious injuries. With spoon-bending powers of optimistic re-framing, Pollyanna manages to see the bright side of being paralysed. Then, in a miraculous feat of recovery, she defies the doctors and learns to walk again. Only a gun-toting homicidal pessimist could stop Pollyanna in her tracks.

Ivan Denisovich Shukhov's deeply religious hut-mate Alyosha is another who takes a perversely positive view of adversity in *One Day in the Life of Ivan Denisovich*, Aleksandr Solzhenitsyn's harrowing tale of life in one of Stalin's labour camps. Alyosha cannot comprehend why anyone should crave release from the appalling rigours of forced labour in the frozen wastes of northern Kazakhstan:

> 'Why d'you want freedom? In freedom your last grain of faith will be choked with weeds. You should rejoice that you're in prison. Here you have time to think about your soul.'

Indeed, after eight years in the camps Ivan Denisovich himself is no longer sure whether he wants to be free or not. In reconciling himself to his incarceration he has grown strangely content with the daily horrors of his existence and now fears change.

Probably the best known of all fictional optimists is the philosopher Dr Pangloss in Voltaire's *Candide*. The well-intentioned but naive young Candide leaves the safety of his castle home in Westphalia and embarks upon a series of extraordinary and frequently disastrous adventures around the world. Throughout his travels Candide is reminded of the maxim of his respected tutor, Dr Pangloss, that 'all is for the best in this best of all possible worlds'. This universally reassuring world-view helps to sustain Candide and Pangloss through the many ghastly fates that await them.

The apparent wisdom of the Panglossian outlook is underlined when the heroes meet Venetian nobleman Seignior Pococurante, a man who is surrounded by beauty and luxury yet finds fault with everything:

> 'Oh what an extraordinary man!' said Candide, muttering to himself: 'what a great genius is this Pococurante! nothing can please him.'

The danger of adopting such a wildly optimistic stance as Pangloss or Pollyanna is that it will shatter in the face of reality. The dazzling light of unquestioning optimism can so easily be replaced by the unremitting gloom of deep pessimism. Someone, presumably a pessimist, once defined a pessimist as an optimist in possession of the facts. Thus did Candide lose faith in the end:

> After having endured a prodigious number of kicks on the backside, of stripes across my shoulders, of strokes with a bull's pizzle on the soles of my feet; after having felt an earthquake; after having been present at the hanging of Doctor Pangloss, and lately seen him burned alive; after having been plundered by order of the divan, and drubbed by a company of philosophers; notwithstanding all this, I believed that all was for the best; but I am now entirely undeceived.

The biology of stress

The fundamental purpose of the stress response is to enable organisms, including you and me, to cope swiftly and effectively with life-threatening challenges. It does this by making us ready for immediate action. Strategically, the short-term stress response – better known as the fight-or-flight response – involves a rapid switch of priorities from long-term to short-term survival. Biological resources are channelled to systems that might be needed to cope with imminent challenges. Whether you must run away or stand and fight, your body will need extra energy, and fast.

Many of the biological changes that accompany the stress response are designed to mobilize the body's fuel reserves, convert them into a form suitable for immediate use and transport that fuel, together with the extra oxygen required to burn it, to the organs most likely to need it – notably the brain and major muscles. This is done at the expense of other biological systems such as growth and reproduction which, although essential in the longer term, are not vital for immediate survival. The hormonal systems that regulate growth and reproduction are plumbed into the stress response and are profoundly influenced by it. Prolonged stress inhibits the secretion of growth hormone and the sex hormones. After all, what use is libido or a fine stature if you have just been mangled by a sabre-toothed tiger or malleted by an assailant?

The manifestations of these biological reactions to stress are perfectly encapsulated in this short passage taken from Pat Barker's 1995 Booker Prize-winning novel *The Ghost Road*. It describes the sensations of a First World War soldier as he waits to go into battle:

> The usual symptoms: dry mouth, sweaty palms, pounding heart, irritable bladder, cold feet. What a brutally accurate term 'cold feet' was. Though 'shitting yourself' – the other brutally accurate term – did *not* apply. He'd been glugging Tincture of Opium all day . . .

What is going on beneath the surface during a stress response? The short-term stress response is mediated primarily by a part of the nervous system called the sympathetic nervous system, which deals with the body's housekeeping functions under normal circumstances and is therefore well placed for rapidly re-adjusting the priorities. The pulse, blood pressure and breathing rate all increase to boost the supply of available energy – hence the pounding heart. In addition to beating faster, the heart pumps a greater quantity of blood with each beat, while the bronchial tubes dilate to allow extra air through with each breath. The blood vessels supplying the muscles expand. The palms of the hands and soles of the feet start to perspire, because a moist surface provides a better grip (the stress response evolved to cope in a world without shoes). The pupils dilate to let in more light and improve vision. Mental alertness is improved and reaction times speed up.

If things get really terrifying the parasympathetic nervous system can trigger involuntary urination and defecation. Besmirching yourself in this way might be messy, but having an empty bladder and bowel could be helpful if things get desperate. It makes you lighter and, perhaps, somewhat less appetizing to a potential predator.

In the meantime, biological functions that are not vital for short-term survival are shut down. Long-term energy reserves in the form of stored fat are broken down into fatty acids and glycerol which can be metabolized immediately. Carbohydrates stored in the liver are mobilized by converting them into glucose. Blood is shunted away from the extremities towards the muscles, heart and brain, and the peripheral blood vessels constrict – hence the cold hands and feet. (When a person gets 'cold feet' in anticipation of an unpleasant event they are showing textbook signs of a stress response.) Energy-consuming digestive processes, including the production of saliva, are shut down – hence the dry mouth, loss of appetite and churning guts.

In ancient China police interrogators were reputed to have exploited the dry-mouth effect of stress to identify lawbreakers. Suspects would be made to fill their mouths with dry, cooked rice. The guilty party would be dry-mouthed from stress and therefore unable to swallow the rice. This use of a physiological response to assess a person's mental state was, in effect, an early version of the polygraph (and probably about as accurate). Churning guts are another classic symptom of the short-term stress response. When someone 'gets the wind up' they are displaying the physiological effects of sympathetic nervous system arousal. Sexual functions are also put on hold until more conducive circumstances return – hence the drooping libido that typically accompanies stress.

Measured in terms of aiding survival in a hostile and dangerous world, the stress response is obviously a Good Thing. A physiologically aroused organism is better able to deal with life-or-death situations. The stress response, in other words, is entirely normal; humans and other species have evolved to respond in this way to stressors. There is nothing odd or maladaptive about finding threats to our wellbeing unpleasant and seeking to avoid them. Whether the stress response remains a Good Thing if it is triggered several times a day, day in day out, under evolutionarily unusual circumstances is another matter, as we shall see later.

Stress has psychological consequences as well as psychological causes. We are back to our two-way street again. Stress changes the way we perceive the world: it affects our senses, memory, judgement and behaviour. For example, the way in which the nervous system processes sensory information is modified by the stress hormone cortisol. A high cortisol level (characteristic of stress) is accompanied by a reduction in sensory acuity (that is, the ability to detect very weak stimuli) and a concomitant improvement in sensory discrimination, allowing us to make finer distinctions between disparate stimuli.[2] All the senses, including taste, smell, hearing and balance, are affected. Thus someone with a high cortisol level may not be able to detect the presence of, say, a very faint sound, but their ability to tell two slightly different sounds apart will be enhanced.

Whatever its evolutionary origins might be, this readjustment in sensory abilities makes good biological sense. When confronted with a sabre-toothed tiger or hostile adversary our Stone Age forebears' chances of survival would have improved if their senses were optimized for dealing with the immediate threat and ignoring anything extraneous. This usually means being able to make fine discriminations between relatively intense stimuli rather than detecting the presence of very faint stimuli. Firm evidence has emerged that another of the hormones released during the stress response – namely, noradrenaline – enhances the signal-processing capability of our sensory systems.

The psychological and behavioural manifestations of stress are grimly relayed in Saul Bellow's *Seize the Day*. It tells the story of a day in the life of Tommy Wilhelm, a middle-aged man on the brink of despair.

Wilhelm has big problems and the world is closing in on him. He has failed as an actor, failed as a salesman, failed as a husband and failed as a son. He lives alone in a New York hotel room, his wife refuses to divorce him, he has no job, no money and his rich father will not help. All in all, things are bad. Wilhelm can see no escape from his troubles and feels he 'would be crushed if he stumbled'. As his personal relationships disintegrate and his debts mount, he finds himself under intolerable pressure – the tell-tale signs of which are noticed by his father:

> Why the devil can't he stand still when we're talking? He's either hoisting his pants up and down by the pockets or jittering with his feet. A regular mountain of tics he's getting to be.

Wilhelm finds it almost impossible to make a decision, and when he does it is usually the wrong one. He lacks concentration and is constantly distracted. Inwardly, he feels as though he is about to burst; he is choked up and has difficulty catching his breath. When Wilhelm argues on the phone with his estranged wife his symptoms grow worse:

> He struck the wall again, this time with his knuckles, and he had scarcely enough air in his lungs to speak in a whisper, because his heart pushed upward with a frightful pressure.

In his rage, he smashes the phone and grinds his teeth – though he is by nature a soft-hearted and sensitive man who would never knowingly cause harm.

Wilhelm displays the self-destructive behaviour patterns of a man at breaking point. His eyes are 'red-rimmed from excessive smoking'; he sits alone in his hotel room, swigging gin from a coffee mug; he consumes drugs to ward off his misery; he cannot sleep, despite the pills. He neglects to wash or shave himself properly and his room is a pit of squalor. Even mundane activities like eating breakfast turn into a struggle against despair:

> Such thinking . . . would grow into a fit of passion if he allowed it to continue. Therefore he stopped talking and began to eat . . . A faint grime was left by his fingers on the white of the egg after he had picked away the shell.

What are the biological mechanisms underlying the stress response? When a stressor provokes a stress response in a person (or a rat for that matter) a complex array of biological mechanisms swing into action, involving myriad nerve connections and chemical messengers. The details are complex and I shall try not to deluge you with them. Certain principles are, however, crucial.

The stress response originates in, and is co-ordinated by, the brain.

Before a stressor can evoke a stress response the organism must first perceive a threat to its wellbeing. This may involve all manner of conscious and unconscious thoughts, beliefs, memories and emotions. Many different parts of the brain play a role in processing all of this information, including the cerebral cortex and other so-called higher centres. Conscious thought need not be involved.

When your brain decides (consciously or unconsciously) that all is not well, the hypothalamus is activated. This region of the forebrain is the source of many of the primary electrical and chemical signals which trigger the full stress response in the rest of your body. (The hypothalamus normally spends its time helping to regulate more pleasurable functions such as eating, drinking and sex.) Elsewhere in the brain, the reticular activating system will be boosting your general level of arousal and awareness, making you more responsive to signals from your sensory organs and less attentive to information of no immediate relevance, such as minor pains or bodily sensations. You do not want to be distracted by an itchy nose when a catastrophe is about to befall you.

The two main biological systems involved in mediating the stress response are the sympathetic nervous system and the hypothalamus-pituitary-adrenal system. The former links the brain to the other internal organs and carries the housekeeping messages needed to regulate essential functions such as breathing, heart rate and digestion. The normal role of the sympathetic nervous system is to mediate the unconscious regulation of basic bodily functions. In a stressful situation it is also the chief mediator of the body's immediate alarm reaction – the so-called fight-or-flight response.

During the initial moments of a stress response the hypothalamus stimulates nerve endings in the sympathetic nervous system and the adrenal glands, causing them to release the hormones noradrenaline and adrenaline.[3] A mildly stressful activity, such as public speaking, will typically elicit a 50 per cent increase in the amount of noradrenaline circulating in the bloodstream and will double the level of adrenaline. For a laboratory rat, being handled or having the cage door opened can elicit an eight-fold increase in adrenaline levels. Humans suffering from chronic stress or anxiety register persistently raised levels of adrenaline and noradrenaline.

Now we turn to the second key player in the stress response: the hypothalamus-pituitary-adrenal system. When this system is activated a cascade of information emanates from the brain. The hypothalamus sends electrical and chemical messages (such as corticotrophin-releasing hormone, or CRH) to the pituitary gland, a pea-sized outgrowth of the brain situated just below the hypothalamus. CRH stimulates the anterior or frontal section of the pituitary to release a second hormone called corticotrophin (alias adrenocorticotrophic hormone, or ACTH). The bloodstream rapidly carries ACTH to the adrenal glands, where it stimulates the outer section of each gland (the adrenal cortex) to release an array of hormones. These hormones in turn act on other parts of the body, modifying the way they work. The whole process is controlled by an elaborate network of feedback and feedforward loops which normally ensure that the response does not get out of hand.[4]

For our purposes, the most important hormones secreted by the adrenal cortex when the hypothalamus-pituitary-adrenal system is activated are the glucocorticoids, a class of steroid hormones akin to the steroids used by doctors for treating inflammation and allergies. In humans the principal glucocorticoid hormone is called cortisol.[5] Cortisol's chief purpose is to help convert the body's energy reserves into a form suitable for immediate use, but it is of special interest here because of its effects on the immune system, a subject we shall examine shortly.

Much of the biological research on stress over the past fifty years has focused on the stress-related activation of the sympathetic nervous system and the hypothalamus-pituitary-adrenal system. Indeed, for some scientists in this field 'stress' is synonymous with the release of hormones by these two systems. According to this operational definition, anyone who has elevated levels of 'stress hormones' (principally adrenaline, noradrenaline or cortisol) is exhibiting a stress response.

In its entirety, the stress response involves a flood of electrical and chemical signals culminating in the release of a cocktail containing over thirty chemical messengers. These include adrenaline, noradrenaline, cortisol, endorphins, encephalins, melanocyte-stimulating hormone, thyroxine, thyrotrophin, vasopressin, aldosterone, renin, growth hormone, glucagon, prolactin, parathyroid hormone, calcitonin and gastrin. They have a wide range of biological effects on target organs throughout the body, including the heart, blood vessels, gut, lungs, muscles and immune system. Thyroxine, for example, stimulates the release of energy stored in fat, while aldosterone (amongst other things) raises blood pressure.

The precise pattern of electrical and chemical messages released during the stress response depends on both the nature of the stressor eliciting it and the organism's appraisal of that stressor. There is evidence to suggest that more noradrenaline is produced in situations where the individual can control the stressor, whereas fear or anxiety provoke a greater adrenaline output. So far, however, the hypothesis that each type of psychological stressor has its own, readily identifiable hormonal fingerprint has not proved particularly fruitful.

The hypothalamus-pituitary-adrenal system and sympathetic nervous system work on somewhat different timescales. Activation of the sympathetic nervous system occurs rapidly: as a rule the fight-or-flight response is underway within twenty to thirty seconds of the alarm bells sounding and subsides within an hour or so of the stressor ceasing. The hypothalamus-pituitary-adrenal system takes longer to get going – minutes or hours, rather than seconds – and its effects may persist for days or weeks. A brief stressor can therefore elicit short-term increases in adrenaline and noradrenaline levels without necessarily raising the level of cortisol.

The effects of stress are not limited to adults. The stress response can be observed in newborn babies as well. For instance, the physical stressor of circumcision elicits a three- or four-fold increase in cortisol levels in babies within half an hour of the operation. Moreover, there is a clear association between the behavioural and hormonal components of the response: the greater the baby's behavioural distress, the higher its cortisol levels. Just because a four-day-old baby boy is incapable of taking legal action does not mean that he isn't stressed. So much for those who still assert that cutting off a baby's foreskin is not stressful.

What is the opposite of a stress response? Humour is a strong contender for the title. When we laugh our psychological and emotional state is about as far removed as it can be from the bowel-churning anxiety and po-faced solemnity of stress. A humorous attitude can often prevent a stressful situation from arising in the first place. As the eighteenth-century English poet Matthew Green put it:

> Fling but a stone, the giant dies.
> Laugh and be well.

Scientists have investigated what goes on inside the body when someone is having a good laugh and, lo and behold, have discovered that the biological changes are virtually the reverse of those that occur during a stress response. When American scientists showed volunteers an amusing video and then measured their blood hormone levels they found that cortisol and adrenaline levels had dropped significantly.

STRESS, IMMUNITY AND HEALTH

The impact of psychological stress on health is no longer an unrecognized and neglected problem. On the contrary, the notion that psychological stress increases our vulnerability to disease now permeates the popular media. If anything, the health hazards are in danger of being debased by hyperbole. We are led to believe that the merest whiff of stress can prostrate us with illness. It is, of course, more complicated than that. But first, how does stress do what it does to us?

We saw earlier that a major element in the stress response is the activation of the hypothalamus-pituitary-adrenal system, triggering the release of cortisol from the adrenal glands. It has long been recognized that glucocorticoid hormones such as cortisol have a profound influence on the immune system. In the mid nineteenth century Thomas Addison, the British physician and pioneering endocrinologist, observed that patients with abnormally low levels of glucocorticoid hormones (a disorder which came to be known as Addison's Disease) had an unusually high number of white cells in their blood. In the 1930s, Hans Selyé discovered that when rats were subjected to severe physical stressors there were visible changes in their immune tissues. Prolonged stress made the thymus shrivel up, while the adrenal glands swelled.

By the late 1940s scientists had established that glucocorticoid hormones can reduce inflammation and suppress various aspects of immune function. The steroids that resulted from this research became the wonder drugs of the 1950s, and they continue to have many clinical applications in the treatment of inflammatory disorders, allergies and autoimmune diseases such as rheumatoid arthritis.

Glucocorticoid hormones like cortisol act upon the immune system in a number of ways. They alter the distribution and movement (or 'traffic') of white blood cells within the body. By stimulating the removal of white blood cells from the bloodstream they reduce the number in circulation and inhibit white blood cells from accumulating at sites of inflammation. They block the production of new lymphocytes in the thymus (which is why the thymus shrinks in animals subjected to severe stress). They can cause the selective destruction (or programmed cell death) of lymphocytes, a process known as apoptosis. They can also suppress the production and release of cytokines, inflammatory mediators, neurohormones and other chemical messengers involved in regulating the immune system, thus rendering lymphocytes less responsive to stimulation.

Several of these effects work at the genetic level. Most cells of the immune system have special receptor sites on their surfaces to which glucocorticoid hormone molecules bind. When this happens the hormone changes the expression of genes within the immune cell and thereby alters its biological activity.[6]

By no means all stress-induced suppression of immunity is mediated by glucocorticoid hormones, however. Stress elicits changes in the levels of various chemical messengers and some of these influence the immune system. In particular, immune function is affected by adrenaline and noradrenaline, which are released when the sympathetic nervous system is activated. There is a firm link between the amounts of adrenaline and noradrenaline secreted in response to a stressor and the degree of fluctuation in immune function.

Because the hypothalamus-pituitary-adrenal system is slower to respond to provocation than the sympathetic nervous system, it is possible for a stressor of short duration to trigger the release of adrenaline and noradrenaline without there being a corresponding rise in cortisol. But the speed and ease with which the hypothalamus-pituitary-adrenal system responds to stressors varies from one person to the next. Research has revealed that some people are particularly sensitive to brief stressors and exhibit unusually large increases in hormone levels, heart rate and blood pressure. These 'high reactors' appear to have extra-responsive hypothalamus-pituitary-adrenal systems. They display an increase in cortisol in response to brief stressors which would not activate the release of cortisol in most people.

Intriguingly, there is evidence from work on other species that individuals who are unusually reactive to stressors tend to be more vulnerable to drug addiction. One study found that mice whose hypothalamus-pituitary-adrenal systems showed the biggest response to a relatively mild stressor (such as being exposed to a novel environment) had a higher risk of becoming addicted to amphetamine when given the opportunity to administer the drug to themselves. This finding fits in with the general hypothesis that addiction to mood-altering drugs, including alcohol, is partly dependent upon how reactive the individual is to environmental stressors.

Excessive activation of the hypothalamus-pituitary-adrenal system and sympathetic nervous system has been shown to have adverse consequences beyond suppression of immunity. Links have been established with certain forms of clinical depression, panic disorders and even alcohol addiction. In fact, any serious disturbance in the complex mechanisms regulating these systems can have serious repercussions for physical and mental health, whether through under-activity or over-activity. There is evidence that unusually low levels of activity in the sympathetic nervous system and hypothalamus-pituitary-adrenal system are associated with a range of complex disorders, including atypical depression, seasonal affective disorder (SAD), post-traumatic stress disorder (PTSD) and, possibly, chronic fatigue syndrome.

Another group of chemical messengers that play an important role in stress-related changes in immune function are the endogenous opioids. These peptide molecules, produced naturally in the brain and elsewhere in the body, are chemically similar to opium-derived drugs like morphine and heroin. They can be divided into two main categories: endorphins (such as beta-endorphin, which is released by the pituitary gland) and encephalins (such as met-encephalin, released by the adrenals).

Among the most useful properties of endogenous opioids is their ability to reduce the perception of pain. Severe stress – of the kind that accompanies serious physical trauma or battlefield injuries – triggers the release of endogenous opioids which have pain-relieving actions akin to high doses of morphine. This has the merciful effect of making the victim practically impervious to pain, a phenomenon known as stress-induced analgesia. The effects of endogenous opioids are not entirely beneficial, however, for 'opioid stress' can suppress the immune activity of natural killer cells and lymphocytes, and hence increase susceptibility to tumours.

Suppressing the immune system is not the only way in which stress can directly damage our health. The mobilization of the body's fuel reserves – a key element in the stress response – entails the rapid release of free fatty acids into the bloodstream. These substances contribute to the build-up of harmful fatty deposits on the inside walls of arteries, a process which leads to coronary heart disease. We shall be looking more closely at this issue in chapter 8, when we consider what the mind can do to the heart.

The quality of stress

What makes a stressor stressful? Although stressors come in all shapes, colours and sizes, it is possible to divine general patterns among the endless variety.

Clearly, some stressors are inherently more severe than others. Losing your spouse or your job, for example, will elicit a bigger stress response than queuing in a supermarket or misplacing a sock. An equally unsurprising conclusion to emerge from research is that intense stressors typically elicit larger changes in hormone levels and immune function than less intense stressors of the same type. However, a number of less obvious generalizations have also come to light.

Stressors share certain basic features which have a substantial bearing on the magnitude of any resultant stress response. Besides intensity, the qualities of a stressor that help to determine its biological impact are its duration (is the stressor brief or long-lasting?); its timing (when does the stressor occur in relation to other biologically significant events such as exposure to disease-causing agents?); its predictability (to what extent can it be anticipated?); and its controllability (can you do anything to alter or eliminate the stressor?). Other things being equal – which, of course, they never are – a stressor has a bigger impact if it arrives without warning, lasts a long time and you have no control over it.

First let us consider duration. The relationship between the duration of a stressor and its biological impact is not as obvious as might be assumed. A stressor can be a brief, discrete event, like making a parachute jump or taking an exam. It can be a prolonged series of intermittent events, such as having sexual difficulties with your partner or aggressive encounters with your boss. Or it can be virtually continuous and long-lasting, such as a chronic, severe illness or the aftermath of bereavement, divorce or redundancy.

Thanks to the capacity of the human mind to dwell upon past horrors and imagine future horrors to come, it is often difficult to determine when a stressor begins and ends. If you are faced with an important exam in six months' time, when do you start to be stressed? Almost certainly long before the actual day of the exam. And when does the stress end? If you performed very badly in the exam you might still be reliving its horrors and getting upset weeks later. A long-term study of caregivers who were looking after a relative with progressive dementia revealed that the stress continued long after the dementia patient had died. Caregivers were still exhibiting symptoms of depression over a year and a half later. As the researchers put it in the subtitle of their paper, 'Chronic stress isn't over when it's over'.

The distinction between short-lived (acute) stressors and prolonged (chronic) stressors is especially prone to become blurred when social stressors are involved. When a person is under stress because, say, their marriage is breaking up, it is difficult to determine whether they are experiencing lots of brief stressors or one long stressor.

Despite these complexities it is nonetheless possible to divine general patterns in the mass of experimental data. By and large, the common sense prediction is borne out: long-lasting stressors do tend to inflict greater damage than brief (but otherwise similar) stressors.

In fact, brief stressors can have the opposite effect on immune function to prolonged stressors. A brief, relatively mild stressor often elicits a transient *increase* in certain aspects of immune function, such as the number of circulating T-lymphocytes, and the quantity and activity of natural killer cells. Prolonged stressors, on the other hand, tend to have a uniformly suppressive effect on the immune system.

Although they tend to be overlooked by the popular media, the immune-enhancing effects of brief stressors have been well documented by psychoneuroimmunologists. In one experiment, for example, volunteers who were exposed to a brief laboratory stressor (mental arithmetic accompanied by loud noise) showed transient increases in the number of suppressor/cytotoxic T-lymphocytes and natural killer cell activity. Similarly, German medical researchers found that parachutists making their first jump exhibited a transient increase in the number and activity of their natural killer cells.

One reason why the effects of a stressor depend partly on its duration is that the stress response is composed of elements which react at different speeds. We saw earlier that whereas the sympathetic nervous system can be activated by a stressor within seconds and stimulate the release of adrenaline and noradrenaline, the hypothalamus-pituitary-adrenal system takes much longer to start generating extra cortisol.

It might be assumed that organisms adapt to prolonged stressors, in much the same way that they adapt or habituate to long-lasting stimuli. But this is not always so. In the previous chapter we saw several examples of how people and animals do not adapt to long-term stress. The immune function of people caring for Alzheimer's victims showed no signs of adapting. Even after years of unremitting exposure to this stressor, the caregivers continued to have poorer immune function and psychological wellbeing. Similarly, neighbours of the damaged Three Mile Island nuclear reactor continued to display stress-related deficits in immune function, altered hormone levels, higher blood pressure and greater psychological distress for years after the accident. Laboratory experiments have produced comparable findings. For instance, exposing rodents to months of a mild stressor (in the form of shifts in the light-dark cycle) produces a prolonged reduction in lymphocyte responsiveness.

Prolonged stressors have been known to evoke complex changes in immune function. In the 1970s Andrew Monjan and Michael Collector of Johns Hopkins University in Baltimore conducted an experiment in which they exposed mice to daily sessions of a noise stressor over a period of five and a half weeks. For the first few weeks the stressor induced the expected reduction in immune function (a drop in the responsiveness and cell-killing ability of B- and T-lymphocytes). This decline in immune function was accompanied by an increase in cortisol. Then, after about three weeks of continual noise stress, the picture changed: immune function recovered to its normal levels and then rose above that of unstressed control animals, before eventually dropping back to normal after five weeks of stress. This experiment demonstrated that the biological effects of a continual stressor may vary markedly over time, and can in fact reverse.

Next on our list of qualities which determine the impact of a stressor is its timing in relation to other events. Of particular significance here is when the stressor occurs relative to exposure to any disease-causing agents (or pathogens) such as bacteria or viruses. The experimental evidence is complex, but the general picture that has emerged is this. When a stressor occurs just before exposure to pathogens it can enhance immune function and increase the organism's resistance to disease. If, however, the stressor occurs during or soon after exposure to pathogens it can impair immune function and make the organism less resistant to disease.

This variable effect of timing was demonstrated in a number of animal experiments in the 1960s. In one experiment, a sample of mice were stressed by exposing them to loud noise and then inoculated with a virus, while a matching sample were first inoculated and then stressed. The mice that were stressed before they were exposed to viruses were more resistant than normal to infection, whereas those that were stressed after exposure to the viruses were less resistant than normal. Stressor timing has been shown to have similar effects on tumour growth in mice.

The predictability of a stressor can also have a bearing on its biological impact. By and large, being able to anticipate a stressor shortly before it arrives is a Good Thing. Laboratory experiments have shown that stressors are less aversive when they are predictable; for example, if the stressor is reliably preceded by a warning signal. When animals are given the choice between a predictable stressor and an unpredictable stressor they have no hesitation: they prefer the predictable stressor, even if it is more intense. Given the means to choose, rodents will opt to receive an electric shock heralded by a warning signal, even though it is three times stronger and lasts nine times longer than the unpredictable shock.

Predictability also lessens the impact of stressors on immune function. French researchers found that unpredictable electric shocks reduced lymphocyte responsiveness, whereas the same quantity of shocks had no immune-suppressive effect when preceded by a warning signal. Once again, a purely psychological variable – predictability – had a major bearing on the biological repercussions of a stressor.

The drawback with experiments of this type is that they deal with predictability in a rather narrow and unrealistic sense. The certain knowledge that an unpleasant event will occur in the space of a few seconds is one thing; it may help by allowing a breathing space for mental and physical preparation. But suspecting that something unpleasant might be going to happen within the next few weeks or months is altogether a different matter, especially if that knowledge is in doubt.

In real life, major stressors such as bereavement or job loss often have an element of predictability, perhaps weeks or months ahead of the event, but they can rarely be predicted with complete certainty. Worrying about potential disasters that might befall us on some future occasion is a propensity that adds to the sum of human unhappiness, even if it does help us to avoid the odd disaster.

CONTROL, CONTROL AND CONTROL

The financial value of houses is said to depend upon three factors – location, location and location. Much the same can be said about the relationship between stress and control.

The impact of any particular stressor depends to a considerable extent upon the degree to which it can be controlled – that is, the extent to which the person on the receiving end has the power to alter, eliminate or escape from the stressor.

We exert control over stressors by performing any behavioural response which enables us to escape from the stressor, terminate it, or reduce its severity. We run away from the sabre-toothed tiger, fell it with our club or keep it at bay. Control can be exerted mentally, too. We can ignore the stressor, deny its reality or re-frame it so that it no longer seems threatening.

It is hardly surprising that our minds – and those of other species – should be so attuned to a sense of personal control, since control over the immediate environment is vital for most organisms' survival. Control signifies autonomy, mastery and empowerment, that mantra of the 1990s. Lack of control means being a passive victim, carried along by the tide. Control is what you do not possess when your commuter train suddenly stops for no apparent reason, or your car grinds to a halt in a traffic jam and you have no idea how long you will be stuck there. Control is what the elderly and infirm stand to lose when they are put into an institution. Control is what we remove from people when we put them in prison. Control is the missing ingredient when you, the humble citizen, tangle unsuccessfully with the dark, immovable forces of bureaucracy. Think of poor Joseph K. in Franz Kafka's *The Trial* – the flavour of which is conveyed by its opening sentence:

> Someone must have been telling lies about Joseph K., for without having done anything wrong he was arrested one fine morning.

Real-life stressors such as bereavement or unemployment frequently involve a profound loss of personal control over circumstances. Depressives often feel as though they have lost control over their lives, which is tantamount to actually losing it.

The importance of control in determining the impact of stressors has been highlighted in numerous experiments, both on humans and other species. Many of these experiments have employed what is known as a 'yoked' design, in which two people (or two rats) are both subjected to precisely the same stressor at precisely the same times, but only one of the two is given the ability to control that stressor by responding behaviourally.

In one typical experiment human volunteers were exposed to a stressor in the form of loud, intermittent noise. Those experiencing the controllable stressor could turn off the noise by pressing a sequence of buttons. Those experiencing the uncontrollable stressor heard precisely the same noises at the precisely same time (they were 'yoked' to the other subject), but their buttons had no effect. The sole variable between the two was psychological: control.

The uncontrollable stressor elicited a greater stress response than the controllable (but otherwise identical) stressor. When exposed to uncontrollable noise the volunteers experienced more feelings of helplessness, unhappiness, tension, anxiety and depression. Their emotional reactions were mirrored in a heightened physiological response: uncontrollable noise evoked a larger activation of the hypothalamus-pituitary-adrenal and sympathetic nervous systems, resulting in a bigger release of stress hormones. As in many comparable experiments, an uncontrollable stressor magnified changes in mood, hormone levels and nervous activity. Incidentally, if this experiment seems artificial, think how stressful it can be to have noisy neighbours who ignore your pleas for peace – an experience that has on occasion driven people to commit murder or suicide.

The classic experiments on control were conducted in the 1960s and early 1970s by Jay Weiss. He demonstrated that an electric-shock stressor was far more harmful to the health of rats if the rats were unable to control the stressor. Rats who experienced uncontrollable shocks lost weight, developed stomach ulcers and showed clear behavioural signs of distress. Those who received shocks of identical severity but were given the ability to shut off the power by performing a behavioural response such as turning a wheel, jumping on to a platform or pressing a panel, suffered little or no adverse effects.

Uncontrollable stressors generally have a bigger impact on immune function, too. Experiments by Mark Laudenslager and colleagues at the University of Colorado in the early 1980s revealed that a single session of inescapable electric shocks substantially reduced the lymphocyte responsiveness of rats, whereas identical amounts of controllable shocks had no such effect on immune function. Again, the impact of the stressor depended critically on its controllability – a purely psychological variable.

Closer analysis of the human data reveals that what really matters is not so much the actual control the individual has over a stressor, but the degree of control they *believe* they have. Perceptions are all-important. So are expectations: a frustrated desire for control appears to be worse than not having any control at all. This was borne out by research conducted at Yale University. Men who were exposed to stressful loud noise which they perceived to be uncontrollable showed a reduction in their natural killer cell activity, whereas men who believed they could control the noise suffered no such decline in immune function. Moreover, the reduction in immune function was greater in individuals who had a strong desire to be in control.

Being unsure of your ability to control a stressor appears to make matters worse. There is evidence that stressors of uncertain controllability have a bigger impact on subsequent health than stressors which are obviously uncontrollable. This research reinforces the age-old wisdom that we should focus our efforts on the problems we can control and stop fretting over the things we can do nothing about.

The psychological effects of stressors are likewise dependent upon control. Uncontrollable stressors carry a higher risk of provoking psychological depression than equally severe but controllable stressors. Some famous (and grim) animal research has underlined the importance of control. When animals are subjected to severe and uncontrollable stressors they can develop a mental state called learned helplessness, which bears many of the hallmarks of clinical depression in humans.

An animal that has been pushed over the brink into learned helplessness becomes very passive. It stops trying to escape and acts as though it believes there is nothing it can do to evade the stressor –

even if there is. Once it has fallen into a state of learned helplessness the animal finds it more difficult to grasp that certain behavioural responses will bring relief. In effect, it sinks into a slough of despondency from which it no longer attempts to escape.[7] Learned helplessness is accompanied by a general deterioration in physical and mental state. The animal develops stomach ulcers and abandons any attempts at social or sexual behaviour.

Control has multiple ramifications for mental and physical well-being. For instance, it helps patients to deal with chronic pain. A person who feels they are in control of their pain and of their life will be consistently better able to cope with prolonged high levels of pain than one who feels helpless. Research has confirmed that boosting patients' sense of personal control helps in the management of pain. This is an important practical issue for cancer patients and others suffering from chronic pain.

Control also makes sense of work-related stress, a subject we shall be exploring in chapter 7. The most stressful jobs are characterized by a combination of high demands and low levels of personal control – in other words, having little say in how or when to perform tasks.

Finally, I suspect that control might well be the magic ingredient in a number of the popular remedies for stress, such as the various techniques for meditation, relaxation and behavioural therapy. Alternative medicine typically offers the patient a greater sense of personal control over his or her problem than conventional medical treatments, which could explain its broad appeal. What you are doing to alleviate your stress may be less important than the fact that you are doing something – anything – and thereby taking control of your predicament. Perhaps we should not agonize over which stress-management technique we use. The best thing may be simply to get on with it and enjoy the benefits of being in control.

The joy of stress

Sweet are the uses of adversity
William Shakespeare, *As You Like It* (1599)

J'aime les sensations fortes.
Ian Fleming, *From Russia With Love* (1957)

Stress has an unremittingly negative image. We are conditioned by the media to associate stress with anxiety, grey hairs, ulcers, unhappiness, all manner of diseases and early death. In our attempts to escape its evil clutches we puff on cigarettes, swig alcohol, swallow pills, nibble chocolate, listen to music, play sport, watch TV, meditate or pray. We may attend expensive courses on stress-management techniques or read books instructing us how to banish stress from our lives. In public relations terms, stress is a disaster area. Yet this unalloyedly negative image of stress is, to say the least, unbalanced.

Sweeping prejudice aside and looking more calmly at the scientific evidence reveals that stress is by no means all bad. Relatively mild, brief and controllable stress can be stimulating and enjoyable. Hans Selyé recognized this over half a century ago; he coined a special name for the beneficial forms of stress, which he dubbed 'eustress'. The more enlightened self-help books now preach the potential benefits of stress, working from the reasonable premise that an optimal level of controllable stress can make people healthier and happier.

Within limits, we perform better when we are mildly stressed. Such short-term improvements in performance are associated with the release of adrenaline and noradrenaline, two hormones which help to prepare organisms for action and improve their ability to cope with immediate challenges. Thus among students taking an academic exam, those who exhibit the largest increases in adrenaline tend to get the best exam results. Similar correlations between stress response and performance were recorded in an extensive study of trainee Norwegian paratroopers. Those who displayed the biggest increases in adrenaline and noradrenaline when they made a training jump also performed best when they progressed to jumping from an aircraft and in written tests of technical competence. In addition, stress hormones dropped back to their normal levels more rapidly in the better performers.

Performance and stress typically have an inverse U-shaped relationship: extremely high or low levels of stress have a distinctly detrimental effect, but intermediate levels go hand in hand with optimal performance.

In contrast with the beneficial effects of activating the sympathetic nervous system, the release of cortisol from the hypothalamus-pituitary-adrenal system is generally associated with a decline in performance. The Norwegian paratroopers who showed the biggest increases in cortisol during training performed worst when jumping. High cortisol levels were recorded in those who were not coping well with the demands of parachute training and who were most afraid.

A similar picture emerges regarding the relationship between stress and immune function. Mild, short-lived stresses can enhance rather than suppress the immune system. A case in point is the increase in lymphocyte responsiveness which occurs in rats when they are placed in a novel environment. These short-term improvements in immune function are primarily mediated by the increases in adrenaline and noradrenaline that accompany activation of the sympathetic nervous system.

Yet in this context, too, cortisol has a uniformly negative effect. When stress is sufficiently severe or prolonged to trigger a release of cortisol, a decline in immune function usually ensues. If there is such a thing as an ideal stress response then it is probably characterized by a rapid but short-lived increase in adrenaline and noradrenaline with no increase in cortisol.

Exposure to brief, moderate stress can have longer-term benefits as well. In the 1960s scientists found that when baby rats or mice are repeatedly exposed to a mildly stressful experience, such as being handled by a human, they grow up to be more stress-resistant as adults. Animals that are handled when young have consistently lower levels of stress hormones; show smaller and briefer stress responses; and behave more calmly when put in challenging situations. All in all, mild stress in early life seems to make them better able to cope with stress later in life (a comforting thought, perhaps, for parents of small children).[8]

Richard Dienstbier of the University of Nebraska has argued that moderate, intermittent physiological arousal, similar to that occurring in the short-term stress response, can have a toughening effect, analogous to the benefits of repeated exercise. This suggestion is supported by persuasive evidence that frequent biological arousal can improve the ability of the body and mind to cope with stressors, leading to improved performance under stress, greater emotional stability and less damage to health.

If these hypotheses are true, it follows that trying to eliminate all possible sources of stress from your life might not be a wise strategy. It is worth remembering that boredom is a potent stressor. As Nietzsche said: 'Believe me! The secret of reaping the greatest fruitfulness and the greatest enjoyment from life is to *live dangerously*!' Or, as a contemporary stress-management expert would say, it is better to burn out than rust out.

THE STRESS-SEEKERS

It is a peculiar feature of humans that they occasionally choose to place themselves in dangerous or stressful situations, purely for the buzz. Voluntarily risking one's life by parachuting or bungee jumping does induce a stress response and physiological arousal, but the fact that the individual has chosen to experience the stressor and is (at least to some extent) in control of the situation, modifies the response.

Studies of individuals taking part in moderately stressful activities such as learning to parachute have revealed that a stress response normally appears in advance of the actual event. Anticipation is half the battle. Just before someone makes their first ever parachute jump their adrenaline and noradrenaline levels are well above normal, as are their blood pressure and heart rate. They also experience many of the symptoms of the fight-or-flight response: anxiety, dry mouth, churning guts, cold extremities and loss of appetite. If they are very disturbed by the whole situation and genuinely frightened about being hurt, their cortisol level could start to rise as the hypothalamus-pituitary-adrenal system becomes activated.

As soon as the jump is over (and assuming it did not end in death or serious injury) most of these hormonal changes are rapidly reversed and anxiety is replaced by elation – which is presumably why people choose to do such things for fun. German research on novice bungee jumpers found that the euphoria induced by their first jump was accompanied by a large release of beta-endorphin, one of the endogenous opioids. The jumpers who experienced the greatest euphoria were the ones whose beta-endorphin levels increased the most. The fun and the physiology were closely intertwined.

Many fictional characters, and some of their creators, have thrived on danger and excitement. Ernest Hemingway portrayed the attractions of danger in his stories and actively sought it in his own life. In his novel *The Sun Also Rises*, for example, he describes the lure of bullfighting during the fiesta at Pamplona in the 1920s. The excited crowd runs ahead of the enraged bulls, leading them to the ring where they will fight. One man in the crowd stumbles, is badly gored through the back and dies. The dead man was twenty-eight and married with two children, but he was a true *aficionado*. A waiter, who is clearly not an *aficionado*, comments acidly on the man's death:

> All for sport. All for pleasure . . . A big horn wound. All for fun. Just for fun . . . You hear? Muerto. Dead. He's dead. With a horn through him. All for morning fun . . .

Hemingway's own life was characterized by a quest for stimulation. He volunteered for service in the First World War and was badly wounded in 1918 while serving with an ambulance unit on the Italian front. He later worked as a war correspondent during the Greco-Turkish war in 1922, the Spanish Civil War and the Second World War. When not chasing wars, Hemingway was passionately involved with prize-fighting, deep-sea fishing, big-game hunting, bullfighting and even lion taming.

Then, of course, we have the very apotheosis of thrill-seekers – James Bond 007; a man for whom danger and excitement are the stuff of life and for whom any form of domestic predictability would be poison. What better note on which to end this section than the following vignette from *The Man with the Golden Gun*.

Bond is in a sticky situation. He is trapped on a private train whistling through the deserted Jamaican countryside. His only travelling companions are Scaramanga, the deadliest assassin in the world (whom Bond has been ordered to kill), a top KGB operative and an assortment of gun-toting mafiosi. All of them are intent on killing Bond. Not the pleasantest of situations. But Bond does not mind. On the contrary – he has deliberately engineered this confrontation with death and is looking forward to it with relish:

> James Bond smiled grimly to himself. He was feeling happy. He wouldn't have been able to explain the emotion. It was a feeling of being keyed up, wound taut. It was the moment, after twenty passes, when you get a hand you could bet on – not necessarily win, but bet on. He had been after this man for six weeks. Today, this morning perhaps, was to come the pay-off he had been ordered to bring about. It was win or lose . . . The adrenalin coursed into James Bond's bloodstream. His pulse rate began to run a fraction high. He felt it on his wrist. He breathed deeply and slowly to bring it down. He found that he was sitting forward, tensed. He sat back and tried to relax. All of his body relaxed except his right hand. This was in the control of someone else.

6

Other People

> Fellowship is heaven, and lack of fellowship is hell:
> fellowship is life, and lack of fellowship is death . . .
>
> William Morris, *A Dream of John Ball* (1888)

We turn now to a fundamental aspect of human life; one that has a critical bearing on our mental and physical health. This chapter is about our personal relationships with other people: our partner, family, friends, colleagues and strangers. Social relationships are seldom the first things that spring to mind when we think of disease, but they undoubtedly have a major impact. More than twenty years' worth of scientific research has accumulated overwhelming evidence that having strong, supportive relationships is good for your mental and physical health.

Relationships are, however, a two-edged sword. Although they are an essential feature of human life and an indispensable source of succour, they can also be a potent source of stress. Being lonely or socially isolated is bad for your health, but relationships can be a cause of torment. Let us start by looking at their dark side.

Hell is other people? – relationships as stressors

> Hell is other people.
>
> Jean-Paul Sartre, *Huis Clos* (1944)

When they go wrong, social relationships can be powerful and prolonged stressors. Moreover they are stressors that we find it difficult to control or escape from.

Take marriage, for example. That there are links between the state of a person's marriage and the state of their health is now beyond dispute. People who are dissatisfied with their marriages are significantly more prone to depression and ill health. The consequences of marital breakdown spill over into most other aspects of life and often disrupt relationships with children, friends and colleagues.

An investigation by Janice Kiecolt-Glaser, Ronald Glaser and their team at Ohio State University has cast new light on the biological and psychological repercussions of marital conflict. For a start, Kiecolt-Glaser's group established that women who had a poor-quality marriage tended to be in a worse state, both mentally and physically, than women whose marriages were thriving. Wives in poor-quality marriages were more depressed and had lower immune function. Essentially the same was true for men: those in low-quality marriages had poorer immune control over latent herpes viruses and a lower ratio of helper to suppressor T-lymphocytes.

Kiecolt-Glaser's group then investigated the short-term biological consequences of marital conflict. Unobtrusive blood-sampling techniques were used to assess the hormone levels, cardiovascular responses and immune function of ninety newlywed couples. While each couple discussed the state of their marriage the scientists monitored the state of their bodies. All the volunteers, incidentally, were healthy and claimed to be highly satisfied with their marriage.

The data revealed a clear connection between emotional conflict and biological responses. The couples who displayed the greatest hostility towards each other during a half-hour discussion of marital problems had significantly lower scores on several measures of immune function, including natural killer cell activity, lymphocyte responsiveness and immune control over latent herpes viruses. For reasons that are not entirely clear, conflict had a bigger impact on the women's immune function than the men's. The argumentative couples also displayed larger and longer-lasting increases in blood pressure.

These mildly hostile interactions were accompanied by a temporary build-up of the stress hormones adrenaline and noradrenaline. Yet cortisol levels did not increase, indicating that this relatively benign conflict was sufficiently arousing to activate the sympathetic nervous system but not the hypothalamus-pituitary-adrenal system. Other research has shown that people whose sympathetic nervous systems become most activated when they interact with their spouses are much more likely to feel dissatisfied with the state of their marriage years later and much more likely to have poorer health.

The connections between marriage and health are complex and subtle. One long-term American study found that the risk of married men developing heart disease increased if their wives were highly educated. Data collected from middle-aged married couples over a ten-year period in the 1960s and 1970s showed that men whose wives had in excess of twelve years of education were 2.6 times more likely to develop heart disease than men whose wives had only a high school education. What could be going on?

This link between a wife's education and her husband's risk of heart disease only applied to men whose wives had jobs. And the risk to these husbands was far greater if their wives had experienced problems at work, such as being passed over for promotion. One possible explanation, therefore, is that the women's frustration at work spilled over into their marital relationships and the resulting emotional tensions had harmful effects on their husbands' hearts.

Hostile interactions between spouses elicit short-term increases in heart rate and blood pressure which, if constantly repeated, might contribute to high blood pressure and coronary heart disease. In line with this theory, researchers have found that a husband's risk of heart disease is greater if his personality is markedly different from that of his wife. Such personality differences between partners tend to exacerbate emotional conflict, leading to more rows and simmering resentments, and – it would appear – more illness.

Interactions between spouses can elevate their pulse rate and blood pressure even when those interactions are not overtly hostile. For instance, our cardiovascular system is stimulated when we attempt to exert social control over someone else, especially if we are unsuccessful. In one experiment, scientists monitored the hearts of men who were either discussing a problem with their wives in a neutral way or deliberately trying to influence or persuade them. The men who were attempting to exert social control over their wives had bigger increases in blood pressure, both before and during the discussion, than the men who merely talked. Interestingly, when the table was turned and the wives attempted to control their husbands, the women's blood pressure did not rise. (Perhaps the wives found it easier to control their husbands than vice versa. Or perhaps there was less ego at stake. One could speculate endlessly.) Research has revealed that those who have a strong but repressed need to influence others – a characteristic that psychologists refer to as inhibited power motivation – tend to have poorer immune function and a higher incidence of illness.

Our relationships with others can amplify the effects of non-social stressors, making us feel guilty or embarrassed about a problem and thus adding to our stress. When we experience disruptive life events or difficulties at work, our interactions with family members and colleagues may make things worse rather than better.

Humans are not the only animals who have social lives. Other species, too, have evolved the biological and psychological means to deal with complex social relationships. Social interactions have a bearing on their immune function and health as well.

If two rats (or two rhesus monkeys or two birds, for that matter) have an aggressive encounter, which need not involve physical violence, the defeated individual's hormone levels may change substantially and its immune function may be adversely affected. Among animals that live in social groups, such as deer and primates, the outcomes of continual social interactions will influence an individual's social status. When a dominant animal is successfully challenged by a subordinate and loses its dominant position it can end up a social outcast, driven away by the rest of the group. It is not uncommon for an animal ousted in this way to succumb to an infection and die not long after its fall from power.

Scientists from the University of Illinois and Stanford University obtained remarkable insights into the biological consequences of social relationships from a long-term study of wild baboons in East Africa. The subordinate males in a baboon troop are frequently on the receiving end of serious aggression from the dominant male and generally have a pretty hard time. The hypothalamus-pituitary-adrenal systems of these subordinate males are persistently activated, resulting in chronically raised levels of cortisol. At the same time their blood is low in HDL-C (high-density lipoprotein cholesterol). This is bad news for subordinate males, because HDL-C prevents fatty deposits from coating the walls of the coronary arteries. The low levels of HDL-C in subordinate males appear to be associated with the chronic stress arising from their social position.

This work on wild baboons also underlined the biological effects of social turbulence. A social group which the researchers had been studying for a long time acquired a new member: a hostile young adult male who suddenly arrived on the scene and decided to join the group. The main targets for his aggression as he tried to establish his social position were the adult females. The change in the social structure which followed was highly stressful, both for the new male and the rest of the group. Blood samples revealed that the interloper had a high cortisol level (indicating a prolonged stress response) and a low lymphocyte count, together with an exceptionally high level of the male sex hormone testosterone. His cortisol levels were at their highest, and his lymphocyte counts at their lowest, during the period of greatest social turbulence when he was establishing himself in the hierarchy. Other baboons in the group who were frequent victims of the new male's aggression had lower lymphocyte counts than those who escaped his attentions. It was evident that the disruption caused by the immigration of this one individual generated considerable stress and affected the immune systems of all those involved.

Similar phenomena have been observed in laboratory studies of monkeys, apes and other social species. Disrupting an established social group (by reorganizing their living arrangements, for example) can increase the risk of group members developing coronary heart disease.

Man's cruelty to man has been amply mapped out in the annals of literature. William Golding's novel *Rites of Passage* contains a perturbing description of a man being destroyed by social stress. The story is set on a sailing ship during the long, slow passage from England to the Antipodes during the early nineteenth century. It tells of the Reverend Robert Colley, an unlovely and unloved parson who dies of shame after he is subjected to degrading humiliations at the hands of the other passengers and crew.

Nature has not smiled on the Reverend Colley, a physically and socially unattractive man of humble origins. Colley is short, badly dressed and ugly. Though obsequious and eager to please, he somehow manages to elicit dislike or stony indifference. Before long, everyone on the ship is either cold-shouldering Colley or playing cruel tricks on him. When the ship crosses the equator, Colley is publicly humiliated and terrified in a ducking ceremony that goes too far. He is subsequently led into the bowels of the ship and urged to drink by the ostensibly genial crew. But Colley, unused to alcohol, gets blind drunk. He makes a ridiculous spectacle of himself, singing, lurching about half-naked and urinating on deck in front of the passengers and crew. When he sobers up he is mortified with shame. Worse is to come when it emerges that this was not the full depth of his humiliation. While he was blind drunk Colley fellated a young seaman.

Struck dumb with the horror of what he has done, Colley locks himself in his 'hutch' and refuses to budge. All efforts to rouse or communicate with him fall on deaf ears as he lies on his filthy bed for days on end, moving one alarmed officer to proclaim: 'Mr Colley is willing himself to death! . . . I have known it happen among savage peoples. They are able to lie down and die.' And not long afterwards Colley is indeed found dead. The narrator concludes:

> In the not too ample volume of man's knowledge of Man, let this sentence be inserted. Men can die of shame.

Colley's private diary, discovered after his death, records his pitiable distress at society's indifference towards him and his unsatisfied need for human companionship.[1]

Hell is alone? – the harmful effects of isolation

> What is hell?
> Hell is oneself;
> Hell is alone . . .
> T. S. Eliot, *The Cocktail Party* (1950)

> In solitude,
> What happiness? Who can enjoy alone,
> Or all enjoying, what contentment find?
> John Milton, *Paradise Lost* (1667)

Relationships might be a potential source of stress, but what of the reverse side of the coin? What happens if we have no close relationships?

The message that emerges loud and clear from scientific evidence accumulated since the mid 1970s is that having a reasonable quantity and quality of social relationships is essential for mental and physical wellbeing. Loneliness and social isolation can inflict greater damage on our health than social stress. We have seen, for example, that marital conflict is stressful and potentially harmful to health. Nevertheless, the evidence shows that, healthwise, people who are married fare better than those who are unattached.[2] Statistics show that the latter have a significantly higher mortality rate and a higher incidence of accidents, infectious diseases and mental illness than married people of the same age. The statistics also show that, on average, men derive greater mental and physical benefit from marriage than women. But there is a price. Men's health suffers more damage if their marriage is disrupted through bereavement, separation or divorce. Suicide and mental illness are a commoner occurrence among widowers than widows, and their overall mortality rate is higher.

Sartre's line that 'Hell is other people' tells only one side of the story; Bertrand Russell was nearer the mark when he wrote in his *History of Western Philosophy*: 'Man is not a solitary animal, and so long as social life survives, self-realization cannot be the supreme principle of ethics.'

The idea that lonely, withdrawn individuals tend to be unhealthy, both in mind and body, has been around for a long time. In the nineteenth century the pioneering French sociologist Émile Durkheim reported that individuals who were not integrated into society had a higher risk of committing suicide. Twentieth-century research has confirmed this finding and established links between social isolation and the incidence of disease and depression, not to mention reduced life-expectancy. Such links hold true even when other medical risk factors such as smoking, excessive alcohol consumption, obesity, lack of exercise and low socioeconomic status have been taken into account and therefore cannot be dismissed as merely an extraneous statistical artefact.

Furthermore, as medical risk factors go, social isolation is a fairly substantial one. Its impact on health and mortality is comparable to that of high blood pressure, obesity and lack of exercise. Research suggests that social factors can have as much impact on health as smoking.

It is all very well to prove statistically that loners are prone to illness. But which comes first – the social isolation chicken or the illness egg? Does loneliness carry a health risk, or do people become socially isolated because they are already ill? It is, after all, perfectly plausible that a person who is sick might find it difficult to develop or maintain personal relationships as a consequence of their physical condition. Or is it the case that poor health and social isolation are not directly related to each other at all, but are instead both by-products of a third factor, such as a misanthropic outlook on life?

The simplest solution to this conundrum is to monitor the social lives and health of a group of people over a long period in order to discern whether the social factors preceded the poor health or vice versa. Prospective research of this sort has been going on since the 1970s, and it has established that social isolation is usually a cause of illness rather than a consequence.

One of the first large-scale studies to investigate the links between social factors and health was based on a sample of five thousand inhabitants of Alameda County, California. At the beginning of the study in the 1960s the researchers catalogued details of the subjects' social lives, recording whether they were married, the extent of their social contact with relatives and friends, and their involvement with clubs, churches or other social groups. The researchers also recorded other factors which might be relevant to future health, including current state of health, use of medical services, smoking habits, alcohol consumption, exercise and activity patterns, body weight and socioeconomic status. (Social class – or socioeconomic status, to use the technical term – is one of the most powerful statistical predictors of future health.) The subjects' health was then monitored over the ensuing years.

The results were clear. Individuals who were least socially integrated and had fewest social relationships were twice as likely to die during the course of the study as those who had the most social relationships. This correlation between social isolation and mortality held true even when the data had been adjusted to take account of all the other medical risk factors. There was no getting away from it: a poor quality social life went hand in hand with a higher risk of dying prematurely.

Comparable studies based on reassuringly large samples have been carried out in other parts of the USA, Scandinavia and elsewhere in the world. The conclusions have been remarkably consistent. For instance, a six-year study of 17,433 Swedish men and women found that those who had the fewest social interactions and least social relationships had a mortality rate 50 per cent higher than those with rich social lives. Similarly, an American study of middle-aged men reported that mortality rates were lowest among men who gave and received the greatest social support, took part in the most social activities, belonged to clubs and organizations, had lots of friends and were married.

With only minor variations, the link between social isolation and subsequent poor health holds true for both sexes, for all ages, for people living in large cities and small rural communities, and for several countries. The phenomenon is definitely not unique to white, middle-class American men, as cynics have suggested.

The link between social isolation and illness also encompasses a wide range of diseases, including coronary heart disease and cancer. The Alameda County study, for example, found that over a seventeen-year period socially isolated women had a significantly higher risk of developing all types of cancer than women with good relationships.[3] Another study, this time of middle-aged Swedish men, found that socially isolated individuals were significantly more likely to have a heart attack. Loneliness is even a predisposing factor for the reactivation of latent herpes viruses, as we saw in chapter 4.

In addition to increasing the risk of developing a disease in the first place, social isolation can impede recovery from illness. Lonely people take longer to get better and, if the illness is a serious one, they are more likely to die from it. This confirms what good doctors have always known: that one of the best tonics for someone who is ill is to have warm, supportive relationships.

One example comes from British research which looked at a large sample of heart-attack survivors. The results showed that those who had regular contact with family or friends, or were members of a club, organization or religious group, had a better chance of surviving the next three years than those who were socially isolated. Much the same is true for cancer. An analysis of 27,779 cancer patients found that unmarried patients had a 23 per cent greater risk of dying from their cancer than married (but otherwise comparable) patients.

Social support also has a bearing on that most natural of processes, childbirth.[4] On average, women who have a high quantity and quality of social support during pregnancy experience shorter and easier labours, deliver heavier babies in better overall condition and suffer from less postnatal depression.

Experiments in the USA and elsewhere have confirmed that mothers do better during delivery if they are attended by another woman who provides emotional support and conversation. Even this limited form of social support is found to shorten labour and lower the likelihood of stillbirth, induction, fetal distress or other complications. American research has shown that the supportive presence of a spouse, partner or friend during labour reduces the risk of having a caesarean section by an average of 10 per cent and cuts the use of epidural anaesthetics by a massive 75 per cent. One social scientist has estimated that if every American woman had someone to support her during labour, maternity costs in the USA could be reduced by two billion dollars a year.

Supportive social relationships are a Good Thing, so we would expect anything that disturbs those relationships to be a Bad Thing. There is overwhelming evidence that the disruption or dissolution of established social relationships can have damaging effects on our physical and mental health. As we have seen, bereavement, divorce and separation can lead to mental and physical illness, and an increase in the suicide and mortality rates of the partners left behind.

Indeed, marital separation and divorce carry health risks that are, if anything, greater than those associated with bereavement. Persons who are separated or divorced have a considerably increased risk of disease, mental illness and premature death. For a divorced American man, the risk of dying from coronary heart disease is double that of a married man of the same age. (According to the statistics, men who are in the process of divorce or separation are also more likely to be the victim of a criminal assault. You just can't win.)

As with most aspects of our biology and psychology, there is nothing unique about humans. Something analogous to bereavement has been observed in several other species where individuals form long-term relationships. The reaction to bereavement is often a restless searching for the lost mate or companion, together with withdrawal and apathy – in fact, the outward signs of depression. In his book *King Solomon's Ring*, the Nobel Prize-winning zoologist Konrad Lorenz described how a greylag goose was affected by the loss of his mate. The goose became restless and flew ever greater distances looking in places where she might be found.

Social isolation and its corrosive effects form a common theme in fiction, too. It is a central strand, for example, in Thomas Hardy's *The Mayor of Casterbridge*. The downfall of the main protagonist, Michael Henchard, is largely a consequence of his inability to relate successfully to others. He quarrels with his business partner, the business fails and he becomes a poverty-stricken bankrupt. Henchard's impulsive, bad-tempered character alienates those who try to help him after his fall from grace. Crushed and disillusioned, his health deteriorates and he finally dies, alone and rejected by everyone except Abel Whittle, a faithful rustic who owes him a favour. Pinned to Henchard's bed is his will, a last bitter testament to his isolation which is read by Elizabeth-Jane, the woman he once believed to be his daughter:

MICHAEL HENCHARD'S WILL

That Elizabeth-Jane Farfrae be not told of my death, or made to grieve on account of me.

& that I be not bury'd in consecrated ground.

& that no sexton be asked to toll the bell.

& that nobody is wished to see my dead body.

& that no murners walk behind me at my funeral.

& that no flours be planted on my grave.

& that no man remember me.

To this I put my name.

Michael Henchard

The importance of social relationships for mental and physical well-being was keenly appreciated by Dr Samuel Johnson. The great sage suffered from recurrent bouts of melancholy and illness throughout his life. (It is interesting to speculate how closely Johnson's depression and physical maladies might have been inter-related.) Johnson believed he had inherited his 'vile melancholy' from his father. It brought him great tribulations. As James Boswell recorded in his *Life of Samuel Johnson*:

> The 'morbid melancholy', which was lurking in his constitution, and to which we may ascribe those particularities, and that aversion to regular life, which, at a very early period, marked his character, gathered such strength in his twentieth year, as to afflict him in a dreadful manner . . .
>
> . . . in 1729, he felt himself overwhelmed with an horrible hypochondria, with perpetual irritation, fretfulness, and impatience; and with a dejection, gloom, and despair, which made existence misery. From this dismal malady he never afterwards was perfectly relieved

Besides his constitutional gloom, Johnson endured a catalogue of physical maladies. In 1783, the year before Johnson's death, Boswell recorded how his friend's fortitude and patience met with severe trials. Johnson had already suffered a 'stroke of the palsy' and was afflicted with gout. He then developed a tumour of the testicle, though to Johnson's immense relief this receded before surgical intervention was necessary. Towards the end of the year Johnson was seized with 'a spasmodick asthma of such violence, that he was confined to the house' and 'an oppressive and fatal disease, a dropsy'.

In the same year, Johnson recorded his own thoughts about his dismal state of health. In a letter to one of his close friends he referred to his melancholy as his 'black dog' – a term that was later used by Winston Churchill to describe his own periodic fits of depression. Johnson poignantly described how the loneliness that had come with old age had made his black dog harder to cope with:

> The black dog I hope always to resist, and in time to drive, though I am deprived of almost all those that used to help me ... When I rise my breakfast is solitary, the black dog waits to share it, from breakfast to dinner he continues barking, except that Dr Brocklesby for a little keeps him at a distance ... Night comes at last, and some hours of restlessness and confusion bring me again to a day of solitude. What shall exclude the black dog from a habitation like this?

Several years earlier, Boswell had recorded this aphorism from Johnson on the subject of relationships, foreshadowing the loneliness of his old age:

> If a man does not make new acquaintances as he advances through life, he will soon find himself left alone. A man, Sir, should keep his friendship in constant repair.

How does it work?

> It is not so much our friends' help that helps us as the confident knowledge that they will help us.
>
> Epicurus (341–270 BC)

If lonely, socially isolated people are less healthy than those with rich social lives, how do they come to be this way? By what means do our relationships affect our health? You will by now have guessed that relationships impinge upon health in a variety of ways. Our perceptions, behaviour and immune function all play a mediating role.

One way in which relationships can help is by providing an antidote to reduce the impact of stressors. Our psychological response to a stressful situation is diluted when a familiar person is present. Think of the parent accompanying the anxious child into the dentist's surgery, for example. The continued support of a partner or friend can help to cushion us against prolonged stressors as well.

Those who lack social support tend to be exposed to the full harmful effects of bereavement and other stressful life events. One study of parents who had lost a child in war or through accidental death found that bereavement was associated with an increase in the parents' mortality rate over the following ten years – but only among those who were widowed or divorced. Bereaved parents who were cushioned by a close relationship with their spouse were statistically at no greater risk of premature death than parents who had not lost a child. Likewise, a study of young men training to be Navy submariners revealed that stressful life events had a greater impact on the health of those with low levels of social support than on men who enjoyed rich social lives with their friends and relatives.

The short-term buffering effects of social support have been verified by numerous scientific observations. Researchers from Cornell University and New York Hospital conducted a simple experiment in which a succession of volunteers took part in a heated discussion while their heart rate and blood pressure were monitored. In these debates each volunteer was verbally attacked by two members of the research team – a stressful situation which consistently provoked an increase in heart rate and blood pressure. A fourth person was also present during the discussion and either supported the subject or said nothing. Results showed that when the fourth person provided support the subject's cardiovascular reaction was reduced, confirming that social support was able to damp down the effects of the stressor. In another experiment, women were asked to perform mildly stressful tasks in a laboratory setting. When the woman had a friend in the room with her during the tests, her heart rate and blood pressure did not increase as much as when she had to perform alone.

We not only benefit from the presence of a companion when we are stressed, we also experience a stronger desire for companionship. Laboratory experiments with volunteers in the 1950s established that when people are in a stressful situation they feel a greater emotional need to have someone familiar nearby. In the parlance of social psychology, stress makes us more affiliative. (You might regard this as a statement of the obvious, mere common sense. But common sense, as the saying goes, has the curious property of being more correct retrospectively than prospectively. Albert Einstein memorably defined common sense as 'the collection of prejudices acquired by age eighteen.')

In addition to cushioning us against stress, relationships have positive benefits in their own right. They can promote mental and physical wellbeing by influencing our perception, our behaviour and our immune function. For a start, other people can help us form accurate and objective perceptions of our own state of health. If we are lucky, our partner, friends or relatives will steer us away from the worst excesses of hypochondria or, at the other extreme, from dangerous self-neglect and denial. Discussing our anxieties can make us less anxious about trivial symptoms and therefore less prone to seek medical care when there is nothing wrong with us. Other people provide a reference point where we can test out our symptoms and get an informal second opinion. Often, all we really need is reassurance. Socially isolated individuals have no recourse to this reality check and may therefore be driven to the doctor's surgery by unfounded anxieties.

Conversely, other people can give us the necessary encouragement to seek timely medical attention when we are genuinely ill. They can protect us from dangerous self-neglect, denial and complacency. Talking to a friend or partner might prompt someone with a serious disease to consult a doctor at an early stage, instead of waiting until it is too late. Prompt attention can be vital if the disease is life-threatening. The evidence indicates that those with rich social networks tend to seek medical attention at an earlier stage than those who are socially isolated. Social pressures may also provide the necessary spur for patients to comply with the medical advice or treatment they have been given, such as taking their medicine when they are supposed to.

Another way in which personal relationships can enhance health is by promoting healthy behaviour. People with high-quality social lives (which might consist of a single very close relationship) tend to show greater regard for their health and avoid things that could damage it. Those who live alone are more prone to over-indulge in food, alcohol or tobacco. Social isolation can foster self-destructiveness: it's all too easy to give in to temptation and indulge in that extra ice cream, cigarette or glass of whisky. The presence of friends or relations, on the other hand, tends to moderate our potential excesses for fear of incurring social disapproval. Social pressures can encourage us to take positive steps, like quitting smoking, improving our diet, losing weight or taking regular exercise. Studies have found that smokers stand a better chance of succeeding to break the habit when they have good social support, and are likelier to relapse when they lack support. One American sociologist has discovered that individuals who have strong social ties are significantly more likely to wear a seat belt. However, it would be foolish to deny that social pressures can also drive us in the opposite direction, pushing us into unhealthy or self-destructive patterns of behaviour.

Obviously, those around us can give help of a purely practical kind, too: providing assistance, money or opportunities to escape from stress. By helping us to focus our thoughts, make the right decisions, buy time or avoid the worst excesses of stress, our friends and relatives can make all the difference when times are hard. A socially isolated individual must cope alone and unaided. To take a mundane example, babysitting grandparents can give hard-pressed parents the chance to recharge their emotional batteries and spend time alone together. And financial support can be vital, especially in countries where the quality of health care depends on a person's ability to pay. One study of women recuperating from surgery for breast cancer found that physical recovery was better among women who had good financial support.

The crucial importance of simple material support provided by a social network is meticulously portrayed in Aleksandr Solzhenitsyn's *One Day in the Life of Ivan Denisovich*. If a prisoner was to survive in the appalling conditions of Stalin's labour camps he had to be a fully integrated member of his labour team, obey its unwritten rules, help his mates and his team leader when needed and be helped by them in return. Each man had to know whom to trust; who would help him and who would refuse; who would save his meagre food ration for him when he was unable to collect it himself; who would let him cadge the stump of a precious cigarette; and who would squeal on him to the camp authorities. Survival depended on the continual trading of small favours. For a man to lose the support of his work team – and especially his team leader – meant almost certain death in the face of Siberian cold, gruelling labour and constant shortage of food.

SOCIAL RELATIONSHIPS AND IMMUNITY

Social relationships can affect our health by more direct means than altering our perceptions or behaviour. There is clear evidence from research on humans and other species that loneliness and social isolation are associated with reductions in immune function and, conversely, that supportive relationships can dilute the immunological impact of stress.

In a series of psychoneuroimmunological studies, Janice Kiecolt-Glaser and her colleagues at Ohio State University uncovered strong links between loneliness or social isolation and impaired immune function. Among samples of reasonably healthy and happy medical students, for example, those who were most lonely achieved significantly lower scores on several measures of immune function.

Specifically, the lonelier students had fewer circulating natural killer cells, less responsive T-lymphocytes, poorer immune control over latent herpes viruses and higher cortisol levels. In addition, it has been shown that individuals with low levels of social support exhibit a weaker immune response when inoculated with a hepatitis-B vaccine.

Similar associations between loneliness or lack of social support and reduced immune function have been unearthed in groups as diverse as the elderly retired, recently-admitted psychiatric inpatients and middle-aged caregivers looking after a chronically sick relative.

There is good evidence that social support can provide a degree of protection against the adverse immunological effects of stress. A study of people who were subject to severe and prolonged stress because their spouses had cancer revealed that those who lacked social support had lower natural killer cell activity and less responsive lymphocytes. (This association was unrelated to depression.) And researchers at the Pittsburgh Cancer Institute found that among women in the early stages of breast cancer, those who received high-quality emotional support from a spouse or other person close to them had higher levels of natural killer cell activity. This finding is potentially important because natural killer cells are believed to play a significant role in preventing the spread of certain tumours.

The disruption of close personal relationships can be particularly harmful to immunity and health. We have already seen that bereavement can result in marked reductions in immune function. Divorce and separation can have similar repercussions.

Kiecolt-Glaser's group found that women whose marriages had broken up within the previous year scored significantly lower than married women on several measures of immune function. Divorced or separated women had fewer circulating natural killer cells than comparable married women. They also had less responsive lymphocytes, a lower proportion of helper T-lymphocytes in their blood and poorer immune control over latent herpes viruses. The women whose immune systems were worst affected by their marital break-up were those who retained the strongest emotional attachment to their former partner. An analogous picture emerged for men. Separated or divorced men had poorer immune function, felt more distressed and had more illnesses than comparable married men. For men, the health consequences of separation or divorce were, to an extent, dependent on their sense of personal control. Men who had initiated the marital break-up suffered less immune impairment, less psychological distress and fewer illnesses than men whose partners had initiated the separation.

Again, there is nothing unique about humans. The disruption of close relationships can damage immune function in other species as well. The most detailed evidence has come from a series of carefully designed experiments with monkeys by Christopher Coe, Seymour Levine, Martin Reite, Mark Laudenslager and other American researchers. These experiments looked at the biological consequences of temporarily separating an infant monkey from its mother. (The separation normally takes place when the infant is fully weaned and physically self-sufficient, but still emotionally attached to its mother.)

When an infant squirrel monkey or rhesus macaque is separated from its mother, both the mother and infant display clear behavioural signs of distress and the infant emits distinctive isolation calls. Even temporary separation produces measurable changes in the hormone levels and immune function of both individuals. Within half an hour or so of being separated, both mother and infant show elevated levels of cortisol, indicating that stress has activated their hypothalamus-pituitary-adrenal systems. Over the next few hours they appear to calm down and the outward behavioural signs of distress diminish. Inside their bodies, however, it is a different matter, and cortisol levels continue to rise. (One of the interesting facets of these experiments is how external appearances – behaviour – can give a misleading impression of what is going on inside the body.) There are also changes in immune function: the separated mother and infant have fewer circulating lymphocytes, less responsive lymphocytes, lower levels of circulating antibodies and reduced antibody responses. These reductions in immune function are normally reversed when the mother and infant are reunited.

Social support can shield the infant monkey against the stress of separation. This phenomenon was demonstrated in a series of experiments in which the infant was housed with familiar individuals after being separated from its mother. When the separated infant was placed in a cage with other infants from its social group, or a familiar adult female (an 'aunt'), the biological impact of separation was substantially reduced. The separated infant had a smaller increase in cortisol and smaller reductions in its immune function, indicating that the presence of familiar individuals lessened the hormonal and immunological effects of separation. Further primate experiments have found that the presence of a friendly, familiar individual can reduce the impact of various chronic stressors on immune function. Things are not so bad when you have a chum with you.

The mother–infant relationship is not the only one that matters, even for infant monkeys. Friends are important, too. Primates who live in stable social groups develop relationships with their fellows, especially those of a similar age. Being separated from these peers can affect immune function. Experiments with juvenile squirrel monkeys and pigtail macaques have found that separating an individual from its peers is stressful and can bring about temporary reductions in lymphocyte responsiveness and other aspects of immune function. As with mother–infant separation, the negative effects of peer separation are diluted by social support.

There is even some evidence that the healthful effects of social interactions can transcend the boundaries between species. Great claims have been made over the years for the putative health benefits of pets (or companion animals, as they are now called by the experts). The notion that animals have healing powers is age-old. The ancient Greeks believed that human diseases could be cured if the afflicted body part was licked by a dog – although they also believed the dogs in question were actually gods in disguise). In Elizabethan times, women were encouraged to keep a lapdog as a prophylactic measure because learned physicians of the day maintained that a small dog clasped to the bosom would absorb diseases emanating from other people. During the 1980s the notion that pets are good for our health was exhumed, dusted off and re-cast in a suitably scientific mould. 'Pet therapy' has now become a growth industry.

Although pet therapists are a soft target for mockery there is tentative evidence that owning a pet might be of benefit to our physical health. When Erika Friedmann and her colleagues at the University of Maryland surveyed a group of patients who had suffered a non-fatal heart attack they found that, other things being equal, those who owned a pet had a marginally better chance of still being alive a year later. James Serpell of Cambridge University found that people who acquired a pet dog experienced a sustained reduction in minor health problems and (not surprisingly) took much more physical exercise. Additional studies have indicated that the companionship of a pet can make it easier for some people to cope psychologically during times of stress.

Being in the presence of a pet is a satisfying, relaxing experience (though not for everyone) and it can induce a physiological relaxation response. Aaron Katcher at the University of Pennsylvania reported that volunteers' blood pressure response to a mild stressor was slightly reduced when their pet was present. It is conceivable that for some individuals the calming, reassuring presence of their pet might help them deal with mild, everyday stressors and thus reduce the risk of their developing high blood pressure.

Pets can act as catalysts for human relationships by facilitating social contacts between strangers. The British zoologist Peter Messent found that pet-owners who were out walking their dogs had significantly more social interactions with other people than those who went out unaccompanied. Indeed, dog-owners did better on that score than adults who were accompanied by small children, which says something rather depressing about British attitudes towards small children. The dog acted as a social ice-breaker; a convenient, unthreatening excuse for strangers to talk to each other.[5]

The lonely future

It is ironic to think that science has recently woken up to the importance of human social relationships for physical health, just as we are losing those relationships. In most industrialized nations the number of socially isolated individuals is rapidly increasing through the effects of social fragmentation, the breakdown of nuclear families and the growing number of old people.

More people are living alone – some from choice, but many because they are separated, divorced, widowed or simply unwanted.

Solitary living is particularly prevalent among the elderly. In Britain, for example, more than half of all those over the age of seventy-five now live on their own. The size of the average household has seen a steady decline throughout this century; it has fallen from more than 4 people per household in 1911 to 2.4 in the early 1990s – and it is still dropping. In the early 1970s one British person in eleven was living alone; twenty years later this figure had risen to one person in seven. If present trends continue, over a third of British homes will be occupied by a solitary person by the year 2016.

Current scientific knowledge suggests that we shall pay a hefty price for the fragmentation of our society – in medical costs as well as human suffering.

7

The Wages of Work

Why should I let the toad *work*
Squat on my life?
Philip Larkin, 'Toads' (1955)

All that matters is love and work.
Sigmund Freud (1856–1939)

And now to work. Like our social relationships with other people, work has two sides. It can be both a source of stress and, conversely, of mental and physical wellbeing. An unsatisfactory job, like an unsatisfactory marriage, can make us miserable and ill. But being unemployed, or worrying about becoming unemployed in the future, can have an even worse impact on our mental and physical health. We shall look first at the dark side of work.

The toad work

Lunch is for wimps
Gordon Gekko, *Wall Street* (1987)

Work-related stress has become a major issue in the eyes of the popular media, the workforce and most enlightened employers.[1] In the USA, work-related stress disorders constitute the fastest-growing category of medical problem. According to one well-researched report, half of all British workers claim they feel stressed and exhausted by the end of a normal working day. Between the mid 1980s and the mid 1990s the number of people receiving counselling for work-related stress doubled. Well over half of all workers now believe their job is the principal cause of stress in their life, whereas a decade ago only a third of workers held this view. What is going on?

Since the early 1980s working practices in industrialized societies have undergone a series of revolutionary changes. The cults of market forces and business school theories have swept away the cosier practices of the 1960s and 1970s, replacing them with a much harsher emphasis on competition, at both the organizational and individual level.

'De-layering' has stripped away middle-management functions and radical 're-engineering' of companies has eliminated many jobs altogether. The majority of employers have indulged in extensive bouts of 'right-sizing', that Orwellian euphemism for sacking people. Jobs that were once in the public sector have been privatized, transforming secure careers into rolling contracts. Meanwhile, those who still have jobs work considerably longer hours. In most Western countries the average hours worked by employees in managerial grades rose by a quarter during the 1980s. More than one in four male employees in Britain now works in excess of forty-eight hours a week. Long working hours inevitably bring conflicts between career and family life, especially for working women.

This Brave New World of work has ushered in an era of widespread insecurity. Far more workers are now either self-employed or in part-time, temporary or casual jobs. In 1971, 18 per cent of British employees were part-time; by 1991 this figure had risen to 26 per cent and it continues to grow rapidly. In certain areas, such as the retail sector, part-time employment is now the norm. For some, the new working patterns offer welcome flexibility, enabling them to balance the conflicting demands of job, family and outside interests. Others, however, find that new working patterns mean insecurity and uncertainty.

The once comforting concept of a secure job for life is virtually extinct. The average person can now expect to have several jobs during their lifetime. Many will spend their working lives as peripheral workers rather than salaried employees, engaged on short-term contracts. We shall all be consultants before long.

According to recent research, over a third of British office workers feel their jobs are not secure – a higher percentage, incidentally, than elsewhere in Europe. In a 1995 survey, 41 per cent of those sampled said they were worried about losing their job within the next year. Their fears are not irrational. According to official figures, at least 80 per cent of British workers personally know someone who has been made redundant within the past few years. (And, as the AIDS epidemic has shown, personal experience is much more effective at spreading fear and changing behaviour than mere statistics. It is one thing to read the cold statistics about AIDS or unemployment, but that cannot compare with the impact of witnessing the effects first-hand.) Lack of job security and fear of unemployment drive those with jobs to work longer, harder and more competitively, though not always more productively.

In short, the modern working environment has become increasingly stressful. Changes in working patterns, combined with recession and unemployment, have meant two things for the majority of workers: greater demands and more uncertainty. Put together, these two ingredients make stress. It hardly needs scientific proof, but research has confirmed that there is a close link between job security and psychological wellbeing. Many people now lack both.

As well as making people unhappy, work-related stress is bad for their health and therefore no friend to the company balance sheets. Employees who are stressed by their job are more prone to accidents and absenteeism, and are less productive. According to well-researched estimates as much as 60 per cent of absenteeism is stress-related. Organizations are now beginning to recognize that stress can have as much impact on their employees' health as exposing them to hazardous machinery.

The economic impact is enormous. Calculating the financial consequences of work-related stress and illness is a tricky business and by no means an exact science; estimates vary widely according to who is doing the estimating and the methods they use. Despite their variability, the figures all point to the same conclusion – that work-related stress constitutes a significant economic and social problem in industrialized society.

Consider this selection of statistics. In the USA the economic costs of stress – measured in terms of absenteeism, lost productivity and health care – have been estimated at up to $150 billion a year. The International Labour Organization calculates that stress costs the UK around 1 per cent of its Gross National Product. According to one set of official figures, stress-related illness loses the British economy an estimated 80 million working days each year, at an annual cost of around £7 billion. A more cautious estimate from the Department of Health put the cost of absenteeism resulting from stress and mental problems at over £5 billion a year. Either way, it is a lot of money – enough to build a few thousand new schools each year. Experts have calculated that stress-related problems can reduce a company's profits by as much as 5–10 per cent. Much the same is true in other industrialized nations.

Even the British law, which is not renowned for climbing on trendy bandwagons, has formally recognized the reality of work-related stress. In a landmark case in November 1994, the High Court ruled that an employer – Northumberland County Council – was in breach of its duty of care to provide a reasonably safe working environment. One of its social work managers, a Mr John Walker, had suffered two nervous breakdowns after he was swamped with an ever-increasing workload of difficult child-abuse cases with no extra resources or support. The unfortunate Mr Walker suffered the classic symptoms of prolonged, uncontrollable stress: exhaustion, anxiety, insomnia, headaches, emotional instability and irritability. This was the first time a British court had held an employer liable for 'mental' (as opposed to 'physical') injury arising from the working environment.

A lucrative new field of litigation has opened up for British lawyers, as workers line up to sue their employers for work-related stress. While lawyers rub their hands in anticipation, insurance companies brace themselves for an epidemic of costly lawsuits. In America, where litigation over work-related stress is already an established tradition, one in seven of all illness compensation claims is now stress-related. And company accountants will be keenly aware that the average sum paid out in compensation for stress-related claims is approximately double that awarded for physical injuries.

Perhaps the clearest proof that work-related stress is now taken seriously is the amount companies are investing in programmes to control stress among their employees. In Britain, over a quarter of employees now work for organizations which take action to counteract stress, such as providing counselling, stress-management training or Employee Assistance Programmes.

Hard-nosed profit-making organizations do not engage in this sort of investment simply to make their employees feel loved and wanted. They do it because it makes sound economic sense. It is estimated that companies save an average of £5 for every £1 they spend on stress-management programmes, mainly through reduced sickness absenteeism and improved productivity. When the predicted tidal wave of spectacular compensation claims reaches the shore, the economic attractions of controlling stress will be self-evident to all.

WHO SUFFERS AND WHY?

Who are the victims of this epidemic of work-related stress? We are all familiar with the stereotype of the highly stressed, power-breakfasting businessman burning the candle at both ends. Top executives are happy to remind us of their gruelling workloads, enormous burdens of responsibility and frenetic schedules (especially when their gargantuan remunerations packages are under scrutiny). It might therefore be natural to assume that our captains of industry suffer more than the rest of us from the damaging effects of stress. Surely the punishing demands placed on these colossi must put them at greater risk from stress, illness and premature death than the proles on the factory floor or the humble functionaries in the accounts department?

Not so. In fact, quite the opposite is true. Life, as irritating pub philosophers are wont to say, is not fair. Research has consistently found that the higher a person's occupational grade or status (and, therefore, the higher their pay) the better their mental health, physical health and longevity. The galling truth is that the fat cats are less likely to get sick or die prematurely than those lower down the greasy pole. Sorry, but there is not an ounce of *Schadenfreude* to be had here.

One of the first scientific studies to uncover the positive correlation between job grade and health was published in the 1960s and drew on data from over a quarter of a million employees of the Bell Telephone Company. The data revealed that the higher an individual ranked in the organizational hierarchy, the less likely they were to suffer from coronary heart disease. To some extent this association between seniority and health was linked to education: those with least heart disease tended to be the highly educated ones, who also tended to be the most senior. Employees with a university education had 30 per cent fewer heart problems than non-graduates. Education alone, however, could not fully explain the association between job status and health. Even after adjustments to the data to take account of educational differences, those on the lowest rungs of the company ladder were still twice as likely to suffer from serious heart disease as senior staff.

Similar conclusions emerged from a classic study of heart disease and mortality patterns among more than 17,000 middle-aged British civil servants in the 1970s. The data revealed that civil servants in the lowest grade band were 3.6 times more likely to die prematurely from heart disease than those in the upper echelons. In fact, a low-grade job was a stronger predictor of premature death from heart disease than those archetypical risk factors smoking, obesity and high blood pressure.[2]

The same basic association between job grade and health has been found in a variety of settings. It holds true for virtually every major illness and cause of death. Moreover, it cannot be explained away by economic or class-related differences in medical risk factors such as smoking, diet or exercise. People of higher socioeconomic status do tend, on average, to smoke less, eat healthier food and take more exercise – but not enough to account for the spectacular health advantages they enjoy. Job grade has a significant bearing on mental and physical health in its own right. Why should this be?

One possible explanation is that employees in the lower strata are unhealthier because their work exposes them to more or worse stressors than those at the top. This seems counter-intuitive, but put aside any prejudices you may have about what makes a job stressful. Stress is not just a matter of having lots of frightfully important, expensive things to do, or working impressively long hours. As we saw in chapter 5, psychological stress arises when the perceived demands exceed the perceived capacity to cope. Jobs can be stressful because there is simply too much to do in the time available, but they can also be stressful because the job-holder lacks the requisite knowledge, skills, training or motivation to do the job properly. Stress is not solely a characteristic of the job; it also depends on the characteristics of the job-holder.

Individuals who reach the senior ranks of their profession are (in theory, if not always in practice) more capable of dealing with larger demands, by dint of experience if not natural ability. They have already been selected, in part, for their ability to cope with the difficulties, demands and stressors of their work. You are unlikely to make it to the upper echelons if you crumple under the normal pressures of work.

Another vital ingredient is control – or, rather, the lack of it. The stressfulness of a situation is intensified when there is no escape or the individual has no control over the stressor. By and large, subordinates have less control over their immediate work environment than those further up the hierarchy. In a poorly managed organization employees may be uncertain about their precise responsibilities and the standards expected of them, which exacerbates their sense of helplessness.

Researchers at St George's Hospital Medical School in London conducted an experiment which highlighted the importance of personal control in work-related stress. Healthy, middle-aged volunteers were asked to perform challenging tasks, such as drawing mirror images of objects, in a laboratory setting. The subjects either had to perform the tasks according to an imposed schedule – a situation in which they had little control – or they were allowed to perform the tasks at their own pace. Subjects who had to perform according to an externally-determined schedule had much larger stress-related increases in heart rate and blood pressure than those who were allowed to set their own work pace. Lack of control heightened the degree of stress and amplified their cardiovascular response. (And for any management consultants who may be reading this, the subjects who were allowed to work at their own pace performed the tasks with greater accuracy.)

Lack of control consistently emerges as a prime cause of job dissatisfaction, and it offers an explanation as to why the hoi polloi suffer more than the chiefs from stress-related problems at work. Giving workers a greater sense of personal control is often an effective way of moderating stress and improving productivity.

It has been suggested that lack of control may play a hitherto unrecognized role in 'sick building syndrome', that medical *cause célèbre* of the 1980s. A large proportion of those who work in so-called sick buildings suffer from persistent headaches, lethargy, stuffed-up noses, watering eyes and other physical complaints. Productivity can fall as much as 40 per cent below normal and absenteeism becomes rife. It has on occasion proved necessary to close down such buildings because they were so unbearable to work in.

Sick building syndrome has traditionally been blamed either on physical attributes of the building, such as bad ventilation and fumes given off by synthetic materials, or on the supposed hypochondria, hysteria and work-shyness of the staff. But research has shown that the physical environment, whilst important, is not the only thing that matters. Psychological factors such as control also feature in the equation. The atmosphere might be objectively cleaner and of optimal temperature and humidity in an air-conditioned office, but the occupants have no control over their environment when the windows are permanently sealed. Given the choice, most people prefer to work in an office where they can open the windows even if, according to objective measurements, their environment becomes smellier, noisier and more polluted as a result.

If having too much work to do is bad, and having no control is bad, then a combination of the two is positively dire. Research has established that the most stressful jobs are those which combine high levels of demand – a large workload and not enough time to deal with it – and a low level of control – insufficient say in how or when to perform tasks. To use the technical term, jobs which combine high demands with low control are said to have a high level of job strain.

The evidence shows that workers in high-strain jobs have a greater probability of suffering from high blood pressure, coronary heart disease and heart attacks. For instance, one study looked at a large number of apparently healthy men employed in a variety of jobs in New York. Unintrusive blood-pressure monitors were used to measure the men's blood pressure throughout the normal working day. The data showed that men in high-strain jobs were three times as likely to have high blood pressure as men in low-strain jobs, even allowing for risk factors such as age, weight, smoking and physical activity. The same holds true for women.

So much for the problem, what about the cure? What can be done to alleviate stress in the workplace? Lots of organizations and employers are making genuine efforts to alleviate the problems of work-related stress. As yet, however, they tend to adopt a rather one-sided view of the problem. Work-related stress stems as much from faults in the organization as faults in the workforce. Many employees who are stressed by their jobs suffer because they are badly managed or inadequately trained. Nonetheless, the usual remedies involve making employees more stress-resistant rather than removing the sources of their stress. The standard stress-management programmes run by most organizations are aimed at teaching employees how to cope with pressure, whether through effective time management or deep-breathing exercises.

In this respect, psychological stressors are still treated in a fundamentally different way from other workplace hazards. The law dictates that employers must eliminate the dangers posed by physical hazards such as toxic fumes and dangerous machinery. Where psychological stressors are concerned, employers are still free to 'cure' the workers rather than eliminate the hazard.

What of work-related stress in literature? I am tempted to quote from *A Christmas Carol*, but I shall leave Ebenezer Scrooge, Bob Cratchit and Tiny Tim to one side. Arthur Miller's play *Death of a Salesman* is a more recognizably modern depiction of a man cracking up under the pressure of failure at work and the dissolution of his own private American Dream.

Willy Loman, a travelling salesman from Brooklyn, has been on the road for thirty-five years. He projects himself to the world as a great success. But things are not as he claims. He has been reduced to working on a commission-only basis instead of a steady salary. Although now in his sixties, Loman must endure the continual pressures of a punishing schedule and impossible sales targets. He is unable to cover his debts, despite working ten or twelve hours a day. His is a classic high-strain job, with large demands and little control.

Mentally and physically exhausted, Loman is prey to memory lapses and suffers from intrusive daydreams about his past. In this condition he is of no use to the business, so his callous boss fires him. The final payment is due on the Loman family home and Willy comes to the conclusion that his life insurance makes him more valuable dead than alive. He ends his final sales trip in a suicidal car crash.

The demands of Willy's working environment and his lack of personal control over his fate are summed up in this epitaph:

> Willy was a salesman. And for a salesman, there is no rock bottom to the life. He don't put a bolt to a nut, he don't tell you the law or give you medicine. He's a man way out there in the blue, riding on a smile and a shoeshine. And when they start not smiling back – that's an earthquake. And then you get yourself a couple of spots on your hat, and you're finished.

The scourge of unemployment

> Work banishes those three great evils: boredom, vice, and poverty.
>
> Voltaire, *Candide* (1759)

In the previous chapter we saw that, stressful as social relationships may be, their absence can inflict even greater damage on our mental and physical health. Much the same is true for work. If you think having to slave away in a crummy job is bad for your health, rest assured that being unemployed is unhealthier yet. The very threat of unemployment can be enough to harm you.

There is clear, consistent evidence from a variety of sources that being unemployed significantly increases your risks of becoming ill and reduces your life-expectancy. When it comes to physical and mental health, the unemployed fare substantially worse than those with jobs: they have a higher incidence of serious diseases such as lung cancer, heart disease and bronchitis; more psychological problems such as depression, neurosis and chronic anxiety; and much higher rates of suicide and attempted suicide. The correlation between unemployment and poor health is so strong that it can be used by regional planners to estimate future requirements for health-care facilities.

Social scientists and epidemiologists have been aware of an association between unemployment and ill health since the 1930s. Long-term studies in America and Britain after the Second World War confirmed that periods of economic instability and rising unemployment tend to be followed by marked increases in mortality, especially among those sections of society most adversely affected by the economic changes. (These fluctuations are superimposed over a general long-term decline in mortality rates that has occurred during the twentieth century.)

We can see how macroeconomic factors might have a bearing on public health through a straightforward chain of events: recession brings unemployment and instability; unemployment and instability bring stress, both for those who lose their jobs and those who fear the loss of their jobs; and stress brings ill health. Workers who remain in employment are adversely affected, too, because recession makes their work more stressful. When economic conditions get tough employees are expected to work harder and have less say in decision-making – in other words, their job strain increases. They also have to carry the psychological burden of insecurity.

Despite the fading stereotypes, this is not just a problem that afflicts men. Unemployed women are less healthy than women with jobs.[3] What is more, the harmful effects of unemployment extend beyond out-of-work individuals to their immediate families, who have a significantly higher mortality rate and incidence of illness than the partners and children of people with jobs.

A classic study of the links between unemployment and health was conducted in Britain in the 1970s under the auspices of the Office of Population Censuses and Surveys (OPCS). Researchers assessed the health, over a ten-year period, of men who had been unemployed and seeking work at the time of the 1971 census. The data revealed that unemployed men had a mortality rate 20 per cent higher than that of men with jobs. The wives of unemployed men also had a substantially higher mortality rate. Subsequent large-scale studies of unemployed men and women in Britain, Denmark, Finland and other countries have found similar – or bigger – increases in mortality associated with unemployment. The higher mortality rate is consistently found to apply to partners and families of the unemployed, as well as the jobless individuals themselves.

In case you were wondering, we can immediately eliminate an obvious alternative interpretation of these statistics. The strong association between unemployment and mortality does not arise simply because sick people find it harder to get jobs (although this does happen to some extent). Increased illness and mortality among the unemployed is primarily a *consequence* of their unemployment, not a cause. Moreover, this rate of illness and mortality cannot be fully accounted for by differences in other factors which are known to affect physical health, such as age and socioeconomic status. The links between unemployment, disease and mortality are undoubtedly complex and subtle. But one thing is certain: they are not just statistical artefacts.

Unemployment and economic insecurity also go hand in hand with increases in mental illness and suicide. There is a firm correlation between unemployment rates and suicide rates in most developed nations. Since the mid 1970s there has been a consistent and alarming rise in suicide among young men in the UK and other European countries. There are strong indications that this increase is attributable in part to the growth in long-term male unemployment.

Apart from the huge toll in human suffering, ill health associated with unemployment has large economic costs. The unemployed and their families visit their doctors more often, make more use of hospitals and other medical services, and are prescribed more medicines than employed people. One study which looked at the aftermath of a factory closure found that the frequency with which the redundant men consulted their doctors rose by 57 per cent and their visits to hospital rose by 208 per cent. A few years after losing their jobs the men were making four times as many visits to hospital outpatient departments as they had been when they were securely employed. In view of the growing numbers of long-term unemployed in industrialized societies, their greater use of health care has critical economic implications.

Why are the unemployed less healthy than those with secure jobs? What are the biological and psychological mechanisms linking unemployment to disease and premature death? This is a complex issue and it would be all too easy to offer simplistic explanations consistent with one prejudice or another. As usual, a number of mechanisms appear to be involved, including that familiar trinity of perception, behaviour and immunity.

Unemployment, and the profound changes in lifestyle it engenders, tends to increase people's anxiety about health and give them more time in which to act upon their anxieties. A change in perception could account for a proportion of the increase in sickness-related behaviour seen among the unemployed, such as their frequent visits to the doctor. But it cannot account for the unemployed having a greater susceptibility to disease and reduced life-expectancy. It is not simply a question of unemployment fomenting hypochondria.

Behaviour is another mechanism that contributes to ill health among the unemployed. At the crudest level, one reason why the unemployed have a higher mortality rate is because they have a greater likelihood of becoming severely depressed and committing suicide. Suicide is only the tip of the iceberg, however. Beneath it lies a mass of illnesses, ranging from minor infections to cancer and heart disease. The higher suicide rate does not by itself account for the increased mortality rate among the unemployed and their families.

The behavioural changes that accompany unemployment can be more insidious than severe depression and suicide. On average, the unemployed lead unhealthier lifestyles than those with jobs. Long-term studies have found that becoming unemployed can make some individuals more prone to self-destructive habits like smoking, drinking heavily or indulging in an unhealthy diet. For a number of those who are unemployed, smoking cigarettes or drinking alcohol are their sole remaining pleasures in life; they help to ease the immediate distress and boredom. A five-year study of school-leavers in Sweden discovered that youths who joined the ranks of the long-term unemployed had a higher probability of smoking and drinking to excess, using cannabis, developing high blood pressure and being involved in crime than their fortunate peers who obtained jobs. Predictably, the unemployed youths also experienced greater physical and psychological problems.

Experts have frequently highlighted the statistical connection between unemployment and crime, though this connection is often pooh-poohed by politicians, perhaps uneasy at the implication that their economic policies could be a factor in soaring crime figures. Whether or not they commit crime, the unemployed are certainly more likely to be on its receiving end. Evidence from the USA shows that unemployed people are much more at risk of being the victims of violent crime.

Behavioural changes cannot account for all the health differences between the employed and unemployed, however. There are other factors at work beneath the surface. Psychoneuroimmunologists have unearthed substantial evidence that the stress, social isolation and loss of self-esteem that accompany unemployment contribute to poor health by affecting the immune system.

One study, for example, found that the lymphocyte responsiveness of unemployed Swedish women was significantly lower nine months after they lost their job. This change in immune function could not be explained by the crude economic effects of unemployment. Thanks to social security provisions in Sweden, the women continued to receive an income representing around 90 per cent of their original pay. This suggests that the changes in immune function stemmed from the psychological and emotional consequences of unemployment rather than its economic consequences.

Unemployment brings with it a welter of personal, social and financial problems that add to the burden. Losing your job is a stressful life event in its own right. But it sets the stage for a whole succession of additional stressors, such as breakdowns in relationships and financial crises.

For a start, unemployment can be very damaging to personal relationships and family life. The rates of divorce, domestic violence and child abuse are significantly higher among families affected by unemployment. Unemployed couples are twice as likely to divorce as couples where at least one partner has a job. The children of the unemployed have a greater risk of being taken into official care. Unemployment conspires to destroy people's social relationships at a time when they need them most.

It is obvious that unemployment creates financial problems and there is little doubt that these contribute greatly to the stress. Research has confirmed that unemployed people with manageable money problems have better mental and physical health prospects than those who find themselves unemployed and in severe financial difficulties. One survey of out-of-work individuals found that those who had been forced to borrow money within the previous year were twice as likely to suffer from depression as those who had not needed to borrow. Thus unemployment creates a self-perpetuating cycle of deprivation, with new debts adding to existing worries.

To make matters worse, the statistics show that once someone has had one spell of unemployment they are more likely to become unemployed again in the future, even if they do find another job.

Money is not the only issue, however. Jobs contribute to mental and physical wellbeing in various ways besides providing cash. A satisfying job can provide a sense of purpose and boost self-esteem. Many of us define ourselves by what we do for a living. Going to work gives a structure to the day, creates social contact with other people and offers a degree of mental and physical stimulation. Even if we were freed from the financial necessity to work, most of us would probably still be happier and healthier doing some kind of meaningful work than staying at home and doing nothing all day. This is borne out in countries where relatively generous state benefits provide a buffer against the worst financial consequences of unemployment: the unemployed suffer health problems nevertheless.

The impact of psychological and emotional factors on health is highlighted by the discovery that the *fear* of unemployment can be almost as detrimental as job loss itself. It seems that a person's health starts to deteriorate from the point when they believe they are going to lose their job, and not when they actually become unemployed. In the study of factory closure described earlier, the employees started making greater use of health care facilities when their employer announced that the factory would have to close. This happened fully two years before the men were made redundant.

Compelling evidence that anxiety about job security can adversely affect health came from a study by researchers at University College London, which was published in the *British Medical Journal* in late 1995. This study tracked, over a period of several years, the health of a vast sample of British civil servants employed in office jobs in London. During the course of the study the government announced a policy of privatizing large numbers of civil service jobs, starting with one particular department (the first of many). From then on, several hundred staff in this department faced considerable uncertainty about their future employment. Work patterns grew more demanding and professional relationships became strained. By the end of 1994, when their department was fully privatized, over a third of these civil servants no longer had jobs.

The four-year period leading up to privatization was a time of tremendous stress and anxiety for these civil servants, and the stress appeared to have an impact on their health. Compared with their colleagues in departments less threatened by job losses, they reported a consistent decline in their general health and experienced significantly more medical problems. Men were harder hit than women – at least, in terms of physical health. This decline in health could not be accounted for by changes in behaviour, such as an increase in smoking or drinking. If anything, the civil servants facing redundancy had marginally healthier lifestyles, on average, than their colleagues in other departments. A more likely explanation is that their health deteriorated because of psychological stress.

Changes in employment patterns mean that far more people now live with the constant fear of unemployment – a fact that might have something to do with the relentless growth in stress-related problems among employees.

What can be done to alleviate the harmful effects of unemployment on the mental and physical health of unemployed people and their families? Solid evidence about the effectiveness of different forms of therapy is thin on the ground, and what little is known barely takes us beyond a glimpse of the obvious. For example, an American study has made the unremarkable discovery that giving the unemployed a degree of financial security helps to maintain their mental and physical health.

More interestingly, research has shown that social support can dilute the harmful effects of unemployment, especially on mental health. Unemployed individuals who have a high quantity and quality of personal relationships are much less vulnerable to depression, anxiety and physical illness than those forced to cope alone. This is wholly consistent with what we saw in the previous chapter and is another instance of how social relationships can cushion us against stressors. Unfortunately, unemployment also helps to destroy those same social relationships. Once again, it is Catch-22 for the unemployed.

Samuel Johnson was right when he wrote these words to Boswell in 1779:

> If you are idle, be not solitary;
> if you are solitary, be not idle.

Sadly, we do not always have the choice.

8

Sick at Heart

We boil at different degrees
Ralph Waldo Emerson, *Society and Solitude* (1870)

Hearts and minds

In this chapter we shall examine how interactions between the mind and the body impinge upon heart disease, the biggest single cause of death in modern industrialized nations. Heart disease accounts for more than 40 per cent of all deaths in most developed countries. In Britain, for example, almost half of all deaths result from heart disease, as against a quarter from the second biggest killer, cancer. (Infectious diseases, which headed the league table in previous centuries, now claim less than one death in two hundred.) But first, what is heart disease?

To stay alive, our brain, muscles and other body tissues need a constant supply of oxygen and nutrients from the blood. To meet this demand the heart must pump between 5 and 15 litres of blood every minute, at a rate that varies according to current needs. The heart is itself a muscle, and a busy one to boot. More than any other muscle in the body it demands a constant supply of blood.

Interrupting the blood supply to the heart can damage part of the heart muscle through lack of oxygen, causing a heart attack (or myocardial infarction). This, in turn, may cause the heart to stop pumping altogether (cardiac arrest). Disease of the heart or blood vessels can also result in an interruption of the blood supply to the brain, better known as a stroke. In the interests of brevity, I shall use the term heart disease in its general sense, meaning diseases of the heart and its associated blood vessels.

Of all types of heart disease it is coronary heart disease, otherwise known as coronary artery disease, which claims the highest number of victims (accounting for about 80 per cent of all heart disease in industrialized nations). The term in fact covers a variety of disorders affecting the coronary arteries – the blood vessels supplying the heart muscle itself. Thanks to a process called atherosclerosis, the coronary arteries thicken and stiffen, and their internal diameter is reduced by the progressive build-up of scar tissue and fatty deposits (or plaques).

Atherosclerosis is a silent, progressive disorder. It develops gradually over a period of years and initially produces no outward signs that anything is amiss. Atherosclerosis can affect arteries anywhere in the body, but is most serious when it attacks the arteries supplying the heart or brain. If the victim is lucky, he or she will receive warning signs in the form of recurrent chest pains known as angina. The pain indicates that sections of the heart muscle are being deprived of oxygen. For unlucky atherosclerosis sufferers, there will be no warning signs; the disease will first manifest itself in the form of a fatal heart attack or stroke. During the Korean War autopsies on American soldiers killed in action revealed that 77 per cent of them had some degree of atherosclerosis, even though they had all been considered fit and healthy.

Decades of intensive research have revealed a great deal about the nature and origins of heart disease. Various factors are known to increase a person's chances of developing heart disease; these include smoking, high blood pressure, old age, being male, obesity, a family history of heart disease, high levels of cholesterol in the blood,[1] lack of physical exercise, diabetes, unemployment and low socioeconomic status.

The connection between smoking and heart disease is particularly striking. Nicotine in cigarette smoke causes the coronary arteries to constrict and encourages the build-up of plaques and clots. Smoking is the strongest and most clear-cut risk factor for cardiovascular disease. A person who smokes twenty cigarettes a day trebles their risk of coronary heart disease. The aggressive marketing of cigarettes in developing nations has contributed to the spread of heart disease worldwide.

High blood pressure (or hypertension) is also high on the list of risk factors; it encourages atherosclerosis by fostering the build-up of fatty deposits and the creation of small scars in the coronary artery walls. High blood pressure contributes to around 20 per cent of all deaths.

One or more of the main risk factors is present in nine out of ten British adults. Approximately 10–20 per cent of adults over the age of forty have high blood pressure,[2] over 25 per cent are regular smokers and 20 per cent take no exercise. Some people have all the risk factors listed above and others besides. The various risks do not operate independently and cannot be looked at in isolation. Two risk factors acting in concert can generate greater harm than if their individual effects were simply added together. For example, in addition to being a major risk factor in its own right, smoking amplifies the adverse effects of high blood cholesterol. Similarly, obesity increases the risk accruing from high blood pressure.

Although huge amounts of data have accumulated over the decades, the presence of conventional risk factors such as smoking and diet is by no means a perfect predictor of who will die of heart disease. All the conventional risk factors between them can explain only half the overall pattern of heart disease within the population. A person who has several risk factors may be free of heart disease, while someone else who has only a few will be afflicted. It is time for us to consider the role of the mind.

The mind in sudden cardiac death and heart disease

One simple connection between the mind and heart disease should already be apparent. A number of the leading risk factors for heart disease clearly depend on our behaviour. Smoking and physical inactivity are obvious examples. Obesity, too, can be a reflection of certain attitudes and behaviour patterns. Individuals who puff, gorge and idle themselves into an early grave do so because of what is going on in their minds. The mind, however, can influence the heart and blood vessels in more direct ways. Psychological stress can inflict damage on the heart and blood vessels over widely varying timescales. The damage can be insidious, gradually accumulating over many years, or it can be dramatic and instantaneous.

Since time immemorial it has been apparent that strong emotions like rage, fear and grief can precipitate cardiac arrest or apoplexy. Literary case histories abound.

Consider the Earl of Gloucester in Shakespeare's *King Lear*. Gloucester, who has been blinded in retribution for his loyalty to Lear, is reunited with his long-lost son Edgar. He realizes with a shock that his illegitimate son Edmund tricked him into being cruelly unjust to Edgar. This ghastly revelation leaves Gloucester torn between two extremes of emotion: delight at finding Edgar alive, and anguish for the awful wrong he has done him. This emotional maelstrom is too much for Gloucester's cardiovascular system to bear:

> His flaw'd heart –
> Alack, too weak the conflict to support! –
> 'Twixt two extremes of passion, joy and grief,
> Burst smilingly.

Then, of course, hordes of characters have died of a broken heart – though not always for the traditional romantic reasons. *Down and Out in Paris and London*, George Orwell's autobiographical (and therefore possibly true) account of life at the bottom of the heap, features the story of Roucolle, a notorious Parisian miser who is so unspeakably mean that he eats cat food, wears newspaper instead of underwear and makes his own trousers out of sackcloth. Two fraudsters persuade Roucolle to invest six thousand francs in a supposedly sure-fire scheme to smuggle cocaine from France to England. Six thousand francs is a mere fraction of the miser's wealth – he has more than that stuffed in his mattress alone – yet it is 'agony for him to part with a sou'. Nevertheless, after much cajoling his reluctance is overcome and he is persuaded to invest in the scheme. Unfortunately for Roucolle, the cocaine turns out to be face powder and the fraudsters vanish, leaving the miser emotionally devastated:

> . . . poor old Roucolle was utterly broken down. He took to

his bed at once, and all that day and half the night they could hear him thrashing about, mumbling, and sometimes yelling out at the top of his voice:

'Six thousand francs! *Nom de Jésus Christ!* Six thousand francs!'

Three days later he had some kind of stroke, and in a fortnight he was dead – of a broken heart, Charlie said.

How can strong emotions cause the human heart to stop beating? Most victims of sudden cardiac death already have a significant degree of disease, such as atherosclerosis. Psychological stress can then act as the final trigger in a variety of ways.

Excessive or uncoordinated activity in the sympathetic nervous system is a frequent contributory factor in sudden cardiac deaths. It is easy to see why. The moment-by-moment functioning of the heart and blood vessels is controlled by the autonomic nervous system (of which the sympathetic nervous system is part). Nerve signals emanating from the brain constantly adjust the performance of the cardiovascular system to match current demands. Violent emotional reactions such as anger or fear are accompanied by activation of the sympathetic nervous system, resulting in stimulation of the cardiovascular system. This makes itself felt as a sudden acceleration in heart rate. When we experience a shock or strong emotion, our pulse races.

Psychological and emotional reactions can stimulate the cardiovascular system just as much as physical exertion. Acute stress or strong emotion can more than double the heart rate – pushing it up to in excess of 180 beats per minute – and cause a marked rise in blood pressure. Indeed, doctors find that the act of measuring a patient's blood pressure in the surgery can be enough in itself to lead to a rise in pressure, especially if the person is anxious about the result or unused to the procedure.

By themselves, psychologically-induced swings in heart rate and blood pressure do little harm to a normal, healthy heart. But it is a different matter if the heart or coronary arteries are diseased. In someone with coronary heart disease, a sharp increase in sympathetic nervous system activity can produce transient disturbances in the electrical control system which co-ordinates the complex pumping movements of the heart muscle. These disturbances, known as arrhythmias, are the immediate cause of most sudden cardiac deaths. The commonest type of arrhythmia, which accounts for around 70 per cent of sudden cardiac deaths, is known as ventricular fibrillation.

Improved monitoring techniques have enabled scientists to track momentary changes in the hearts of conscious patients, with startling results. We now know that mild, everyday stressors can provoke minor heart malfunctions far more easily than had hitherto been realized. In a patient with coronary heart disease, even something as innocuous as performing difficult mental tasks in a laboratory setting is capable of inducing brief interruptions in the blood supply to the heart. Such interruptions can occur without causing chest pains or any other overt symptoms and are known as silent ischaemias. (Ischaemia simply means an inadequate blood supply caused by a partial blockage of the blood vessel feeding that part of the body.)

Most episodes of ischaemia in daily life do not occur during violent physical exertion, as might be supposed, but during relatively sedentary activities. The majority are triggered by psychological and emotional factors rather than muscular activity.

Experiments conducted at Yale University on heart disease sufferers have shown that acute psychological stress is, if anything, an even more potent trigger for silent ischaemia and other malfunctions than pedalling vigorously on an exercise bicycle. In a study conducted by researchers from the University of California at Los Angeles, coronary patients were monitored while they performed stressful tasks like speaking publicly on a personal subject or mental arithmetic. The psychological stressors provoked big increases in arterial blood pressure and abnormal movements in the heart wall in over half the patients. These abnormalities were comparable to those caused by physical exercise and in most cases they resulted in silent ischaemia.[3]

The cardiovascular system is particularly affected by psychological stressors that have personal significance for the individual concerned. Making a person speak in public about themselves or their partner evokes bigger increases in blood pressure than having them perform an abstract mental task. And strong emotions like anger can have dramatic effects on the coronary arteries of cardiac patients.

Researchers at Stanford University found they could induce perceptible constrictions in the coronary arteries of cardiac patients merely by asking them to recall a recent event which had made them angry. (Remember that mixed blessing of the human mind: the ability to be stressed by events in the past or future.)

People with coronary heart disease are especially vulnerable to the effects of psychological stress because diseased coronary arteries do not respond in the normal way to nervous stimulation. In a healthy person, an acute stress response normally causes the coronary arteries to dilate, helping to boost the flow of blood to the heart. Stress can, however, have the opposite effect in patients with coronary artery atherosclerosis; their coronary arteries constrict instead of dilating, thereby reducing the blood supply to the heart just as the demands placed on it rise.

The sharp increase in blood pressure elicited by an acute psychological stressor can occasionally cause a fatty deposit inside a diseased coronary artery to become detached from the artery wall. The detached plaque may block the artery, causing a heart attack, or block a vessel supplying blood to the brain and cause a stroke.

Acute psychological stress can also help to precipitate sudden cardiac death through the partial activation of blood-clotting mechanisms. You will recall that the stress response is designed to prepare the body for possible injury. As part of this pre-emptive defence mechanism, certain components of the blood-clotting process are activated. As long ago as the 1950s researchers discovered that the blood of tax accountants coagulated more quickly at times of the year when their work-related stress levels were especially high. (I shall refrain from commenting on the attractions of research which requires large volumes of tax accountants' blood.) Psychological stress stimulates the platelets, small disc-shaped blood cells which help to halt bleeding, and it affects the levels of fibrinogen, another substance involved in the blood-clotting process.[4]

For a person whose coronary arteries are clogged with fatty deposits, this partial activation of blood-clotting mechanisms can further reduce the flow of blood to the heart or cause the formation of a blood clot (thrombosis), with potentially lethal consequences. To make matters worse, the stress-related activation of blood-clotting mechanisms is more pronounced in heart-disease sufferers than in healthy individuals.

There are several ways, then, in which acute psychological stress can precipitate a cardiac catastrophe in someone with pre-existing heart or coronary artery disease. Not all the effects of stress are so sudden, however. It can contribute to the underlying heart disease in a gradual and insidious way, damaging the heart and blood vessels by degrees over a period of years rather than minutes. How?

Remember what happens during the stress response. Psychological stressors activate the sympathetic nervous system, stimulating the release of adrenaline, noradrenaline and other hormones, and increasing the heart rate and blood pressure. A primary function of the stress response is to mobilize the body's energy reserves for immediate use. Long-term stores are converted into free fatty acids which are released into the bloodstream where they help to form deposits on the coronary artery walls, contributing to atherosclerosis. In addition, the rapid increase in blood pressure which accompanies the stress response can cause minute injuries to the walls of the coronary arteries. The resulting scar tissue then contributes to atherosclerosis.

When the stress response occurs day in, day out over a period of years, these processes can inflict gradual, cumulative harm on the heart and blood vessels.

The environment in which we now live has made the problem worse. The stress response evolved over millions of years to enable us to cope with immediate threats that would normally culminate in fight or flight – in other words, vigorous muscular activity. The initial stress response increases the pulse rate and constricts blood vessels in the skin and internal organs, producing a rapid rise in blood pressure. Intense muscular activity can partially offset this rise in pressure, because the blood vessels supplying the large muscles dilate when those muscles are used. But if the stress response is not accompanied by muscular activity, the rise in blood pressure is unabated. In industrialized societies we rarely respond to psychological stressors with vigorous physical activity. We do not run away or fight – we sit and fume. Our blood pressure therefore stays high.

There is abundant evidence linking psychological factors to the development of hypertension. One long-term study, which tracked a thousand initially healthy subjects over twenty years, discovered that men who displayed high levels of anxiety at the beginning of the study had double the risk of developing high blood pressure in later life.

Psychological stress in its various guises is equally capable of fostering heart disease in other species. For example, the social stress that arises when a stable social group is disrupted accelerates the onset of atherosclerosis in monkeys fed on a high-fat diet. The monkeys who develop the worst atherosclerosis are those whose lives are most stressful – namely the dominant males, who must constantly struggle to maintain their social position, and the subordinate females, who get bullied and harassed all the time. Social stress can cause certain strains of rats to develop chronic high blood pressure.

In another parallel with humans, researchers at the University of Pittsburgh were able to prevent this stress-induced atherosclerosis in monkeys by giving them a beta-blocking drug called propranolol. Beta-blockers are commonly used for treating high blood pressure, angina, abnormal heart rhythms or anxiety in humans. They reduce the activity of the sympathetic nervous system and slow the heart down.

Coronary-prone personalities and heart disease

> But he that hides a dark soul and foul thoughts
> Benighted walks under the mid-day Sun;
> Himself is his own dungeon.
>
> John Milton, *Comus* (1637)

The belief that individuals with certain personality characteristics are especially prone to heart disease is an ancient one. For centuries, there has been speculation about the role of traits like hostility, ambitiousness and impatience in the genesis of heart disease. We are all familiar with the popular stereotype of the heart attack waiting to happen: the stressed, aggressive, competitive, middle-aged businessman. In 1910 the great physician Sir William Osler observed:

> It is not the delicate, neurotic person who is prone to angina, but the robust, the vigorous in mind and body, the keen and ambitious man, the indicator of whose engine is always at full speed ahead.

In the 1930s Franz Alexander, one of the pioneers of psychosomatic medicine, proposed that people with high blood pressure were distinguished by repressed hostility and anger. According to Alexander, who held Freudian views, the repression of these deep-seated emotional impulses produced an inner conflict which manifested itself in raised blood pressure. (The empirical observation was correct even if the theoretical interpretation was off beam.)

Modern research on the links between personality and heart disease was shaped by the seminal work of two American cardiologists, Meyer Friedman and Ray Rosenman. It all started in the 1950s when Friedman and Rosenman noticed that a remarkably high proportion of their cardiac patients shared distinctive personality traits. These men (and they were mostly men) were ambitious, competitive and impatient. Their body movements and speech patterns were abrupt and intense. They loathed having to wait around or waste their time. Friedman and Rosenman noticed that the waiting-room chairs became worn and shiny because these patients were so restless.

Friedman and Rosenman refined their informal observations into a testable theory: that individuals who display this set of behavioural and personality traits have a significantly heightened risk of coronary heart disease. These supposedly coronary-prone characteristics observed by Friedman and Rosenman were dubbed the Type A behaviour pattern.[5] In a preliminary test of this theory (the first of many) Friedman and Rosenman divided a sample of men into two groups, according to whether or not they exhibited Type A characteristics. They then looked for signs of coronary heart disease. The results were consistent with the Type A hypothesis: 28 per cent of Type A men had clinical signs of heart disease, compared with only 4 per cent of the non-Type A men.[6] Friedman and Rosenman's work gave rise to decades of systematic research into the links between heart disease and the Type A behaviour pattern.

THE TYPE A BEHAVIOUR PATTERN

So, what exactly is the Type A behaviour pattern? This is not an easy question to answer and it brings us straight away to one of the chief problems with this whole field of research. Type A behaviour is a broad-brush concept with no precise textbook definition; it means somewhat different things to different scientists and the definition has evolved over the years. Moreover, the Type A category encompasses several personality traits and behaviour patterns. Notwithstanding these difficulties, it is possible to identify the wood in amongst the trees.

The Type A behaviour pattern describes the general behavioural style with which a person responds to their physical and social environment. It has been epitomized as the behaviour of someone locked in an incessant and aggressive struggle to achieve more and more in less and less time. (Do you know anyone who has never felt like this?) Various combinations of over thirty specific behavioural characteristics have been lumped together under the Type A umbrella. This fuzziness about the meaning of Type A has inevitably given rise to inconsistencies in the research data.

Fortunately, there is a degree of consensus about the elements which form the crux of Type A. They are:

- free-floating hostility
- aggressiveness
- competitiveness
- a constant and harrying sense of time urgency
- impatience
- constant striving for ill-defined goals
- vigorous, abrupt patterns of speech and movement

There are many variations on this basic theme, but they all amount to roughly the same thing. Depending how Type A behaviour is defined and measured, and which population is being sampled (white middle-class American men, a random cross-section of all adults, young children, women, elderly Swedes . . .), anything from 15 to 70 per cent of the population are categorized as Type A.

What are you if you are not Type A? With stolid predictability, the antithesis of Type A is called Type B. The Type B personality is trickier to define. In essence it simply means not Type A. The typical Type B person is laid-back, relaxed, unaggressive, easy-going, readily satisfied and not driven by ambition.

One weakness in the Type A/B concept is immediately apparent. It is clearly a gross over-simplification to divide the rich and endlessly complex variation in human personality into just two crude categories. It does not take a professional psychologist to see that there are more than two types of people in the world. A person might well have only a few of the Type A characteristics; there are plenty of impatient individuals who are not particularly ambitious or competitive, for example. And the definition of Type B as 'non Type A' is too vague by far. To be fair, researchers usually divide their subjects into at least four categories: A1 (full Type A), A2 (partial Type A), X (a mixture of A and B) and B. Even so, this is not an enormous improvement.

Like any attempt to divide all of humanity into a few catch-all categories, the Type A/B classification is bound to be inadequate. The medical scientists who pioneered this field of research excelled at the medical side of things, but their psychological formulations fell short of perfection. Yet despite its undoubted crudity the Type A concept has proved to be surprisingly fruitful, as we shall see.

It may be easier to convey the spirit, if not the precise letter, of the Type A concept by giving a selection of illustrations from the world of fiction. There are innumerable Type A personages to choose from. One of the most amusing is Basil Fawlty, the hysterically manic hotel proprietor created by John Cleese in the 1970s TV comedy series *Fawlty Towers*. Fawlty is a Type A to conjure with. All the prime danger signs (except, possibly, competitiveness) are present in an extreme form: intense impatience, appalling irritability, corrosive sarcasm, a jackhammer style of speech, a seething disdain for virtually everyone he meets and a blue touchpaper so short that he explodes into a frenzy at the tiniest provocation. When dealing with his hotel guests Fawlty is, to quote his astonishingly composed wife Sybil, liable to be found 'spitting poison at them like some benzedrine puff-adder'.

In Shakespeare's *The Taming of the Shrew* we have another Type

A in the person of Katharina, 'an irksome, brawling scold' renowned throughout Padua for her bad temper and harsh language.

> Her only fault – and that is faults enough –
> Is that she is intolerable curst,
> And shrewd and froward so beyond all measure . . .

'Curst', by the way, meant bad-tempered; 'shrewd', perverse; and 'froward', wilful or headstrong. The waspish (or do I mean shrewish) Katharina ties up her sister, smashes a lute over her music instructor's head and hits her future husband on their first encounter. In stark contrast, her younger sister Bianca is a mild-mannered, sober, calm, amiable Type B. So stroppy is Katharina, so ready to take offence, so easily roused to anger, so aggressive in her dealings with others, that her father despairs of ever finding her a husband.

But Kate meets her match in husband-to-be Petruchio, a feisty, assertive, quick-tempered man – in Kate's words, 'a mad-brain rudesby, full of spleen'– who is as much of a Type A as she. His fuse is, if anything, even shorter than Kate's: he wrings servants' ears over a trivial misunderstanding, throws food and dishes at them when a meal fails to please and strikes the priest during his wedding ceremony. One of Petruchio's servants, comparing his master to Katharina, comments that 'he is more shrew than she'. Petruchio is undaunted by the prospect of their 'two raging fires' coming together. Their courtship is a process of psychological warfare, at the end of which Petruchio succeeds in breaking Kate's spirit and taming the shrew, thus turning her into a docile wife. (Or does he?)

A latter-day Type A is Lieutenant Colonel Gowrie, the Commanding Officer in Alan Judd's *A Breed of Heroes*. This highly acclaimed novel of modern military life is set during a tour of duty in Northern Ireland during the Troubles. The CO is a fanatic who sets impossible goals for himself and everyone else, driving his men mercilessly. He holds strong views on most matters and will brook no opposition. When he gets worked up about something – which happens all the time – the veins in his temples bulge, his teeth clench and his face turns red. His style of speech, too, is typically Type A: intense, exaggerated and often accompanied by violent, jabbing movements with his finger or swagger stick.[7] Hostility is constantly in the air: 'the personality of the CO pervaded the building and induced in all who entered it a sense of urgency bordering on panic'.

The second-in-command of the battalion, and perfect foil to the CO, is a marvellously amiable Type B called Anthony Hamilton-Smith:

> [Hamilton-Smith] never hurried, never worried and had never been known to be angry. Nor had he ever been known to work. No one knew what he did all day, but it was generally agreed that his presence lent to the battalion a certain tone.

Described by one of his men as the last of the great amateurs, Major Hamilton-Smith has been passed over for promotion, a situation that leaves him unmoved. He is unflappable and serenely detached from the life-and-death dramas that daily unfold around him. Unlike his colleagues, he never appears tired, depressed or irritable, whatever the pressures. He deals with crises in a light-hearted way bordering on flippancy, in marked contrast to his intensely serious CO. In Hamilton-Smith's opinion there are very few things in life that cannot wait until tomorrow.

Type A – the evidence

How strong is the evidence for the putative link between Type A behaviour and heart disease? The Type A concept has fallen under a cloud of doubt in recent years and is now regarded as a promising hypothesis that fell short of everyone's high expectations. But is this a fair judgment?

The initial observations of Friedman and Rosenman indicated that something interesting was going on, but they were not conclusive. Over the following decades a number of major research projects were undertaken to investigate the putative connections between personality and heart disease. These studies typically started with a large sample (often measured in the thousands) of people who were free from heart disease and otherwise healthy. The subjects were assessed for Type A characteristics and medical risk factors such as smoking, high blood cholesterol and obesity. They were then monitored over a period of several years (in some studies, subjects have been tracked for over twenty years) to see who developed heart disease or died from it. The researchers could then determine whether the individuals' personality characteristics were related to the subsequent pattern of heart disease and mortality.

One of the first large-scale investigations was launched in 1960. The sample comprised over three thousand healthy, middle-aged men. At the outset, approximately half of them were found to be Type A. Eight and a half years later the data revealed that the incidence of coronary heart disease was twice as high among Type A men as Type Bs, even when other risk factors such as smoking were taken into account. Type A behaviour emerged as a clear risk factor for heart disease in its own right, comparable in strength to smoking, high blood pressure or high blood cholesterol. Further large-scale studies in the USA and elsewhere have found similar associations between Type A behaviour and the subsequent risk of heart disease.

To reach the firmest possible conclusions about the link between Type A and heart disease, scientists have pooled the results of all published studies on this subject and then analysed them collectively, a procedure known as meta-analysis. When this is done a reliable, if modest, correlation is found between Type A characteristics and coronary heart disease. An authoritative meta-analysis was carried out in the mid 1980s by Stephanie Booth-Kewley and Howard Friedman of the University of California at Riverside, using data from eighty-seven separate studies published in the scientific literature since the 1940s. Booth-Kewley and Friedman's review confirmed that the Type A behaviour pattern is reliably correlated with heart disease and comparable in its impact to other, more conventional risk factors.

By the early 1980s the Type A behaviour pattern had gained official, US government-approved status as a risk factor for heart disease. The Review Panel on Coronary-Prone Behavior, an authoritative group of eminent scientists, decreed in an official report that Type A should be added to the official list of risk factors for coronary heart disease, alongside smoking, high blood cholesterol and high blood pressure.

Much of the early research on Type A looked only at men. Later research, however, demonstrated that the link between Type A behaviour and heart disease holds true for women as well. Type A women, like Type A men, have twice as much chance as Type Bs of suffering heart disease. The proportion of women who exhibit Type A behaviour is comparable to that of men. Type A behaviour – and heart disease, for that matter – should no longer be regarded as a predominantly male affliction.

Complementary evidence has been obtained by examining the consequences of modifying Type A behaviour. If Type A behaviour encourages the development of heart disease then eliminating such behaviour should, logically, lower the risk of heart disease. By and large, this is indeed what happens.

It turns out to be quite feasible to modify Type A behaviour, and the consequences of doing so are consistently beneficial. Teaching individuals to identify and avoid situations in which they exhibit Type A behaviour, to relax and control their anger, can subdue key elements of the behaviour pattern like hostility and impatience.

Modifying Type A behaviour in this way has all sorts of beneficial effects; for example, it reduces blood pressure and blood cholesterol. But the real proof of the pudding is that moderating Type A behaviour in people with coronary heart disease lowers their risk of dying prematurely.

Among the first studies to highlight this effect was the Recurrent Coronary Prevention Project, a long-term survey of over a thousand men who had already suffered at least one non-fatal heart attack. A random selection of these men received special counselling aimed at curbing all aspects of their Type A behaviour. The counselling succeeded in reducing the men's Type A behaviour, especially their hostility and sense of urgency. It also halved their risk of suffering a further heart attack. The fact that reducing Type A behaviour in this way can lower the risk of heart disease strongly implies that Type A behaviour and heart disease are connected.

Curbing Type A behaviour might reduce the risk of heart disease, but is it entirely a good thing? Type A behaviour is widely believed to go hand in hand with greater achievement and success at work, which might explain why organizations tend still to favour Type As when selecting candidates for jobs or promotion. At face value, the competitive, thrusting, ambitious Type A seems a better bet than the lackadaisical Type B.

The notion that Type As achieve more than Type Bs is another of those common-sense assumptions that turns out to be largely untrue. There is precious little evidence that Type A individuals perform better on average than Type Bs. And training employees to moderate their Type A behaviour does not appear to impair their efficiency or effectiveness. This has been confirmed in professions as diverse as salesmen and army officers. If anything, taming Type A excesses increases employees' efficiency and effectiveness, as well as making them less likely to end up in a cardiac ward or a morgue.

So far, then, the Type A theory seems solidly convincing. Several well-conducted scientific studies, monitoring reassuringly huge samples of men and women over long periods, have consistently found significant associations between the Type A behaviour pattern and subsequent heart disease. In addition, researchers and therapists have demonstrated that reducing Type A behaviour leads to an appreciable reduction in the risk of heart disease.

And yet. And yet. Conclusive as all of this may seem, there are uncomfortable inconsistencies in the research data and a shadow of uncertainty now hangs over the Type A concept. A significant number of published studies have failed to find a meaningful link between Type A behaviour and heart disease. The discrepancies have led several authoritative critics to pronounce that the concept of Type A behaviour as a risk factor for heart disease is all but defunct.

In the face of such criticism, defenders of the Type A concept have pointed out that many of the inconsistencies in the data stem from weaknesses in the research methods used, and not weaknesses in the Type A concept itself. At the risk of descending into technicalities, let me outline a few of the main problems.

One difficulty with the research in this field is that several techniques have been used for measuring Type A behaviour. Unfortunately, each of these techniques measures something different. The original and best method for assessing whether a person is Type A is by using a tailor-made interview.[8] The structured interview is expensive and time-consuming, but it is reliable and produces solid data. It is widely regarded as the gold standard. Researchers who have used structured interviews to assess Type A have, on the whole, found statistically significant associations between Type A behaviour, high blood pressure and subsequent heart disease.

Since the 1970s a number of paper-and-pencil methods have emerged as time-saving alternatives to the structured interview.[9] These rely on volunteers providing written answers to standard questions. Their big advantage is that they are easy to administer and allow researchers to study larger samples than would be practicable if each subject had to be interviewed. Unfortunately, these techniques have also proved to be less effective than the structured interview. Studies using paper-and-pencil methods have generally found weaker links (if any) between Type A behaviour and heart disease. This may be as much a product of deficiencies in the measurement tools as any weakness in the actual relationship between Type A behaviour and heart disease. Perhaps it is no coincidence that inconsistencies in the research data started to appear in the late 1970s – at about the time when paper-and-pencil methods came into widespread use.

A second problem is that the impact of Type A behaviour in a healthy person is unlikely to be identical to the impact it would have in a patient with heart disease. Some of the negative findings published in the scientific literature have come from studies of coronary heart disease patients, whereas the subjects of earlier research were for the most part healthy at the time the studies commenced. Type A behaviour might well make little difference to a person with serious coronary heart disease, but this does not mean that it is irrelevant for healthy people.[10]

Notwithstanding the ambiguities in the Type A concept and the undeniable discrepancies in the research data, there is clearly a solid core of truth here. There are good grounds for believing that certain elements of the Type A behaviour pattern do act as risk factors for heart disease in later life. The question is, which ones?

ANGER AND HOSTILITY

Researchers have tried to unpack the portmanteau Type A category to see which of its many components are truly significant. This subtler approach has confirmed that some Type A characteristics play a greater part in fostering heart disease than others. Indeed, a variety of traits that have traditionally been swept under the Type A umbrella may actually be of no consequence. It is now widely agreed that the crucial components of the Type A cluster are hostility and anger.

The idea that people with hostile personalities are more susceptible to heart disease has been floating around in the scientific literature since the 1940s. Since the mid 1980s this link has been repeatedly confirmed by observation.

Hostility and anger consistently correlate with heart disease, poorer general health and increased mortality. This has been true in nearly every study in which they have been measured, including studies that have failed to find any correlation between heart disease and the much broader Type A category. A follow-up study published in 1995 assessed the health of middle-aged women doctors who had graduated from the University of California School of Medicine at San Francisco in the 1960s. Women who had, in their initial assessment, been characterized as having low levels of hostility were in generally better health when middle-aged than their more hostile peers.

A particularly strong association has emerged between heart disease and what psychologists call 'anger-in': the tendency to feel angry but not express this anger openly, even in circumstances where it would be appropriate to do so. One American study which tracked several hundred people over a twenty-two-year period found that individuals who suppressed their anger when they were provoked (for example, in confrontations with their spouse) had a mortality rate 70 per cent higher than those who openly expressed their anger. Accordingly, many experts on stress management now recommend that we should give vent to our spleen – preferably in a constructive, controlled manner – rather than bottle it up. So go on: complain to the waiter, remonstrate with the neighbour whose barking dog keeps you awake at night, write that angry letter to the newspaper, say what you really think of your boss's management skills.

Just as anger and hostility have proved to be critical risk factors for heart disease so, conversely, have other Type A attributes turned out to be largely irrelevant. The well-known 'hurry sickness', which is considered archetypally Type A, has scant bearing on the risk of heart disease. Another traditional component of the Type A cluster that has no clear link with heart disease is drive, or job involvement. In short, the familiar stereotype of the coronary-prone, ambitious workaholic is not borne out by the data. The research paints a different picture of the future coronary heart disease victim: a peevish, belligerent, short-tempered misanthrope.

How does it work?

It is one thing to discover a statistical association between Type A behaviour and subsequent heart disease, but what are the biological mechanisms that underlie this association? How does the Type A behaviour pattern manage to cause long-term damage to the heart and coronary arteries? Not surprisingly, perception, behaviour and physiology all play a role.

Psychological research has revealed wide-ranging disparities between Type As and Type Bs in their attitudes towards health and their sickness-related behaviour. Type As are inclined to notice and complain about minor pains and physical symptoms, though they do not always respond appropriately to their symptoms and are less likely than Type Bs to act prudently if they think they have a genuine illness brewing. For their part, Type Bs tend to make more objective judgements about their state of health. Type Bs who feel ill are wont to take preventative measures, such as avoiding overwork, whereas Type As typically plough on regardless until they keel over. One experiment found that when volunteers were asked to perform demanding physical tasks, like walking on a treadmill, Type As worked harder than Type Bs but were less willing to admit it when they became tired.

Type As are prone to use the psychological defence mechanism of denial. As we have seen, denial can be dangerous if you happen to have a serious disease. Type As who experience chest pains and other warning signs of serious heart problems wait longer on average than Type Bs before seeking medical help. Some of them presumably die as a result of their denial and procrastination.

A more obvious way in which the Type A personality contributes to heart disease is through lifestyle. Research shows that Type As tend to lead unhealthier lives: they get through greater quantities of alcohol, tobacco, and fatty foods than Type Bs. In addition, Type As have a higher statistical risk of being involved in accidents and dying from their injuries. Being competitive, hostile and impatient probably does not help.

Yet behavioural differences like these cannot fully account for the higher incidence of heart disease among Type As. The association between Type A and heart disease holds true even when the statistical effects of other risk factors are eliminated from the analysis. The Type A behaviour pattern is an independent risk factor for heart disease in its own right and cannot be dismissed as a by-product of agents such as nicotine or cholesterol.

BIOLOGICAL REACTIVITY AND THE TYPE A PERSON

We turn now to what goes on beneath the skin – to the biological mechanisms by which Type A behaviour impinges upon the internal workings of our bodies. The principal conclusion that has emerged from a wide array of research is that the hearts and blood vessels of Type As incur damage over the years because Type As are biologically more reactive to stimuli and stressors.

We saw in chapter 5 that humans (and other animals) differ considerably in the way they respond to any particular stressor. This is as true for their cardiovascular system as it is for their hormones or immune system. When a random sample of people are exposed to precisely the same stressors, certain individuals will consistently display larger increases in their heart rate and blood pressure than their fellows. They exhibit greater biological reactivity (or, to be precise, greater cardiovascular reactivity).

Biological reactivity is linked to heart disease. It has long been known that people who show unusually large increases in blood pressure in response to physical stressors (such as immersing a limb in freezing cold water) are more prone to develop coronary heart disease in later years. A similar relationship applies to psychological stressors. A person whose cardiovascular system consistently reacts in an exaggerated way to everyday psychological stressors is at greater risk from heart disease. The same thing happens in other species, too. Monkeys whose heart rate increases the most in response to a psychological stressor, such as the anticipation of capture, are more vulnerable to heart disease than their less reactive peers.

Humans who show strong reactions of this kind tend to have higher blood pressure even when they are resting, and are at greater risk of developing hypertension. One study of adolescents found that individuals who exhibited the biggest cardiovascular responses to a stressful interview also had higher blood pressure readings throughout the normal school day (which is, after all, composed of a succession of stimuli, challenges and minor stressors). Highly reactive individuals are at greater risk from cardiovascular malfunctions such as ischaemia when they are exposed to a mild psychological stressor.

The rot starts young. Individuals differ from one another in biological reactivity from an early age. These differences are fairly stable characteristics and tend to persist through to adulthood. Research has found that children who display unusually large responses to mild stressors have a bigger risk of developing high blood pressure later in life.

Another twist in the tale concerns smoking. Researchers have discovered that highly reactive individuals find it harder to give up smoking. Those whose heart rate and blood pressure increase the most in response to mild psychological stressors have the biggest likelihood of failing in their attempts to quit the habit.

But where does Type A behaviour fit into this story? The basic conclusion from a mass of research is that people with Type A behavioural characteristics tend to be biologically more reactive. Nature has apparently decreed that if you tend to be a seething mass on the outside then you tend to be a seething mass on the inside as well. One reason why Type As are especially prone to heart disease is because their cardiovascular systems are continually reacting more vigorously to the stressors and stimuli of everyday life.

There is a reassuring bulk of experimental evidence to support this hypothesis. It was discovered decades ago that Type A men have higher levels of the stress hormones adrenaline and noradrenaline than Type Bs during the working day, when they are exposed to numerous minor stressors. This hormonal difference between Type

As and Type Bs disappears at night when they are asleep. Subsequent research has established that Type As consistently display larger hormonal, cardiovascular and immune reactions to psychological stressors.

Laboratory experiments have shown that when Type As are exposed to a mild stressor, such as having to perform mental arithmetic, take part in a hostile interview or play a demanding electronic game, they display significantly bigger increases in heart rate, blood pressure and stress hormone levels than Type Bs. Essentially the same thing happens in response to real-life stressors as well. One study found that Type A male medical students had bigger cardiovascular responses than Type Bs when taking an exam. (I confess I was surprised to discover there is such an entity as a Type B male medical student, but apparently they do exist.) Moreover, once the Type A's cardiovascular response has been activated it takes longer to subside after the stressor has ceased. The Type A heart therefore beats harder, and for longer, than the Type B model.

The greater biological reactivity of Type As bites even harder if they already have diseased coronary arteries. Among sufferers of coronary artery disease, those who have the most pronounced Type A traits (especially aggression, anger and hostility) have the greatest likelihood of experiencing a silent cardiac malfunction in response to a relatively mild psychological stressor.

Why should a Type A be more biologically reactive? An obvious explanation is that the attitudes and behaviour of Type A individuals create added opportunities for normal life to upset them. Because of the way they view the world, Type As encounter stress in everyday situations which would not disturb Type Bs. They create rods for their own backs. The competitiveness, ambition and impatience of Type As catapult them into stressful situations, whereupon their free-floating hostility and aggressiveness impel them to react emotionally. Everyday situations like being stuck in a traffic jam or a supermarket queue provoke a more vigorous biological response in the Type A because the Type A is more annoyed. The Type B car driver who is overtaken by another vehicle is sublimely unconcerned, but the Type A driver is enraged because some ******* ****** has cut him up.

A further clue as to why Type A individuals react so vigorously to stressors lies in their need to control situations. A person who tries to exert control even though a situation is inherently uncontrollable will only make matters worse. We saw in chapter 5 that uncontrollable stressors are generally more potent than controllable stressors and that trying, but failing, to achieve control makes them even more stressful. Whereas a typical Type B would recognize defeat at an early stage and gracefully surrender to the uncontrollable stressor, the Type A doggedly continues trying to gain control and thereby aggravates the problem.

The social responses of Type As are also important. Experiments have revealed that people with hostile, aggressive personalities – the hallmarks of Type A – show bigger increases in blood pressure during situations of social conflict. In one experiment, women volunteers were asked to perform a challenging, mildly stressful task while one of the researchers deliberately harassed them. The women whose blood pressure shot up the most in reaction to this irritating experience were the ones who had previously been rated as having the most hostile personalities. (Imagine the irascible Basil Fawlty being forced to draw mirror images and count backwards while a nagging egghead in a white coat stands over him with a stop-watch and criticizes his performance. I mean *really*!)

Social support – or rather, the lack of it – can contribute to the Type A's heightened risk of heart disease. We saw in chapter 6 that people are better equipped to cope with stress if they have supportive personal relationships. Social scientists have made the unsurprising discovery that individuals who behave in a hostile, aggressive way and harbour cynical, disparaging attitudes tend to alienate those around them and consequently receive less social support. Hostility and aggressiveness – the key elements of the Type A behaviour pattern – are hardly endearing characteristics and may corrode the Type A's social relationships.

The conscious attitudes, beliefs and behaviour of Type As are undoubtedly crucial elements in their greater biological reactivity. Yet they may not be the entire story. Some intriguing research suggests that the biological reactivity of Type As could have roots that go deeper than conscious thoughts and attitudes. When scientists monitored patients who were undergoing coronary bypass surgery they found that Type A patients were more reactive than Type Bs even when they were unconscious. The Type A patients exhibited significantly larger increases in systolic blood pressure during the operation, despite being fully anaesthetized.

This discovery implies that there is more to being Type A than just having a particular set of conscious attitudes. One interpretation is that Type A behavioural characteristics – such as aggressiveness, hostility and impatience – might be symptoms of a person's heightened biological reactivity, as well as causes of that reactivity. This suggestion is supported to a degree by the discovery that Type A behaviour can be curbed by giving subjects beta-blocker drugs which reduce their sympathetic nervous system reactivity. Calming the inside calms the outside as well.

But what (I can almost hear you say) about the immune system? Although immunity does not play a starring role in heart disease, it does at least have a cameo part. Individuals whose cardiovascular system is highly reactive to stress also tend to show the biggest changes in immune function. Cardiovascular reactivity and immune reactivity go hand in hand, possibly because they are both mediated by the sympathetic nervous system. When scientists at Carnegie Mellon University in Pittsburgh measured cardiovascular and immune responses to a brief psychological stressor, subjects whose hearts responded most vigorously also showed the biggest changes in immune function. (Their lymphocyte responsiveness fell in response to the stressor and the numbers of circulating suppressor/cytotoxic T-lymphocytes and natural killer cells rose.)

Another study, this time by scientists in the Netherlands, used a laboratory task as a psychological stressor to invoke immunological and cardiovascular reactions in volunteers. Once again, the two reactions went hand in hand. Individuals whose heart rate and blood pressure increased the most in response to the stressor also tended to exhibit the biggest change in immune function (as measured by an increase in the number of circulating natural killer cells).

Evidence like this implies that our immunological and cardiovascular responses to psychological stressors could be governed by the same mechanisms. This could be important, in the light of growing evidence that immunological processes may be involved in the development of atherosclerosis. If so, then Type A reactivity could increase the risk of heart disease via its effects on the immune system, as well as through its more direct effects on the heart and blood vessels.

The connection between personality and heart disease has been at the focus of intensive research spanning four decades. Despite this, it remains surprisingly controversial – surprising, that is, given the weight of solid evidence supporting the connection; surprising also given the less than perfect evidence for some of the more readily accepted risk factors like dietary cholesterol. Again, we are faced with that old double standard: 'physical' causes for disease are inherently easier to believe in than 'psychological' causes.

A cautious analyst would conclude that the old-style Type A behaviour pattern is not as clear-cut and consistent a predictor of heart disease as conventional risk factors like smoking and hypertension. As a concept it may have outlived its usefulness. Nonetheless, there is compelling evidence that key components of the Type A pattern, notably hostility and anger, do play a substantial role in the development of heart disease. These traits can be changed, and changing them can lower the risk of heart disease. Finally, we have several perfectly respectable biological and psychological mechanisms that can explain how the connection between personality and heart disease actually works. There can be little doubt that the manner in which we think and react to the world does have a significant bearing on our risk of developing heart disease.

9

The Mind of the Crab

He who desires but acts not, breeds pestilence.
William Blake, *The Marriage of Heaven and Hell (1790–3)*

Cancer ranks second only to heart disease as a cause of death in industrialized nations. One person in four dies of cancer and one in three of us will develop some form of cancer during our lifetime. About a third of all patients in general hospitals are there because they have cancer. After decades of outstanding biomedical research, our knowledge of cancer exceeds that of any previous generation – yet it is still inadequate. Even now, around half of all cancers that are diagnosed are incurable.

Cancer is a blanket term covering over a hundred diseases with a multiplicity of causes. What they all have in common is some form of malignant growth resulting from an abnormal and uncontrolled replication of cells. In many respects it is astonishing that cancer is not more common. The human body consists of somewhere in the region of 10^{13} (that's 10,000,000,000,000) separate cells,[1] each of which contains a full set of genes and has the inherent capacity to grow and divide, even if that capacity is normally suppressed.[2] Each of those 10^{13} cells therefore has the potential to become cancerous (or neoplastic, to use the technical term) and must be kept under precise and constant control. The relative rarity of cancer, at least prior to middle age, is a testament to the body's superb cellular control mechanisms.

Occasionally, however, the complex regulatory processes within our cells break down and cells start to grow abnormally. If uncontrolled cell division is allowed to continue, a malignant tumour will eventually form. The tumour cells grow and divide rapidly, producing new tissue which invades and destroys the surrounding healthy tissue. In ancient times, the spidery outgrowths spreading from tumours were likened to the legs of a crab, the Latin name for which is *cancer*. And that is how cancer (and this chapter) got its name. Cancerous cells may break off from a tumour, travel to other parts of the body via the bloodstream or lymphatic fluid, and form new cancerous outposts – a process known as metastasis. The development of secondary tumours following metastasis marks a serious escalation in the progress of the disease.

Cancer is predominantly a disease of middle and old age. In comparison with hazards such as accidental injury, it poses only a minor statistical risk to children and young adults. Once we reach late middle-age the incidence rises sharply. The commonest forms of cancer in industrialized societies are those which attack the lung, colon, breast and prostate. The underlying incidence of most cancers has not changed hugely in recent decades, although the absolute number of deaths from cancer has shown a marked increase, primarily because the elderly now live longer and account for a greater proportion of the population. We all have to die of something. In Britain, for example, the total number of patients being treated for cancer has risen by a quarter since the mid 1980s.

Certain types of cancer are on the rise. Foremost among these is lung cancer, which rose dramatically after the Second World War and is now the commonest form of fatal cancer in many industrialized nations. Approximately 90 per cent of lung cancer cases are attributable to smoking. In some countries the incidence of lung cancer among men has begun to fall gradually, reflecting an earlier decline in smoking. However, because of the long time lag of up to thirty or forty years between starting to smoke and developing the disease, the incidence of lung cancer has continued to climb among women, who only started smoking in great numbers after the Second World War. In Britain, for example, it rose by 39 per cent between 1979 and 1990. Smoking is also thought to be partly responsible for an alarming rise in bladder cancer.

Another cancer that is very much on the increase is malignant melanoma, a virulent form of skin cancer.[3] The number of cases in Britain has doubled since the mid 1980s. As with lung cancer, the upsurge in malignant melanoma is, to a degree, attributable to changes in behaviour. In this case it is our fondness for over-exposing our unprotected bodies to bright sunlight. As a modest consolation for the big rises in lung cancer and malignant melanoma, the incidence of stomach cancer has dropped considerably over the past sixty years, probably helped by improvements in diet.

The causes of cancer are complex and a long way from being fully understood. Several factors are known to boost the risk of developing cancer. Smoking and exposure to intense sunlight are well-known examples. And we all know that basking in the ionizing radiation from a leaking nuclear reactor or hydrogen bomb blast is not a good idea. The importance of other putative risk factors, such as certain foodstuffs, environmental pollutants and viruses, is more open to dispute.

As with heart disease, the relationships between the various risk factors for cancer and the eventual disease are complex. A single risk factor may be neither necessary nor sufficient to cause cancer. Smokers can remain free of the disease, while non-smokers can fall prey to it. Individuals may be especially vulnerable to certain risk factors and not others. We differ in our sensitivity to sunlight, for example. Furthermore, particular combinations of risk factors can have a disproportionately powerful impact, the whole being greater than the sum of the parts. What role does the mind play in this complex causal nexus?

The mind in cancer

Cancer frightens people, even though it is often curable and claims fewer victims than heart disease. It carries menacing connotations of uncertainty, pain and prolonged suffering, followed inevitably by an undignified death. Attitudes have changed in recent years as the public's level of understanding has increased, but cancer still has the capacity to terrify.

The notion that psychological and emotional states such as grief, depression or stress can foster the development of cancer has been around for centuries. Nearly two thousand years ago, the Greek physician Galen declared that melancholic women were more likely to develop cancer. This idea was echoed in the eighteenth and nineteenth centuries, when various prominent physicians noted that melancholics who had experienced grief or loss seemed vulnerable to cancer. In the 1890s the British physician Herbert Snow observed that a surprisingly high proportion of the cancer patients he saw had lived difficult lives and had recently experienced a traumatic event, such as the death of a loved one.

Nowadays few scientists would accept the crude proposition that a mental state can independently cause cancer by conjuring up a tumour out of nowhere without any intervention from other causal agents. Instead, scientific attention has focused on two somewhat more refined propositions. The first is that aspects of our personality or psychological state may modify our susceptibility to cancer, perhaps by making us vulnerable to cancer-causing influences or by impairing the body's ability to detect and destroy new cancers. The second idea is that psychological factors affect the chances of survival and recovery of people who already have cancer. As we shall see, there is considerable scientific evidence to support both hypotheses. Let us examine a small sample of this evidence.[4]

The termination of a close relationship, whether through death, divorce or other means, has often been linked with cancer. When American psychologist Lawrence LeShan looked at more than four hundred cancer patients he discovered that a remarkably high proportion (72 per cent) had suffered the loss of someone close to them not long before the onset of their cancer. In contrast, only 10 per cent of comparable people without cancer had suffered such a loss. The implication was that the loss of a loved one had heightened the risk of cancer.

Another study found that women with breast cancer had experienced a greater tally of life events and emotional losses before their cancer was diagnosed than cancer-free women who were generally comparable in other respects (such as age, marital status, number of children and social class). Stressful life events also appeared to affect their prognosis: those women who had endured the most life events in the year preceding the onset of cancer had the least chance of surviving the disease.

Retrospective evidence like this must be treated with caution, however. Human memory is highly imperfect and those who are ill often need to attribute their illness to a specific event in their life. For definitive evidence we must turn to prospective studies, in which subjects' psychological states are assessed before rather than after their cancer is diagnosed. There have been numerous studies of this type and most have unearthed significant connections between psychological factors and subsequent disease.

In a study conducted in the 1970s, 110 men with suspected – but as yet unconfirmed – lung tumours underwent psychological assessment. Each was later diagnosed as having either lung cancer or a benign tumour. The results showed that men who had recently experienced psychological stressors such as the end of a relationship, fear of unemployment or an unstable marriage, were more likely to have a malignant tumour than a benign tumour. In fact, psychological measures were as effective at predicting which individuals would have lung cancer as their history of smoking.

Another study investigated the incidence of cancer among medical students who had graduated from Johns Hopkins University between 1948 and 1964. Each subject was assessed, both medically and psychologically, at around the time of graduation. Their health was then monitored periodically over the ensuing years. It emerged that graduates who felt they had not experienced close relationships with their parents, especially their fathers, had a greater likelihood of developing cancer in later years. A follow-up investigation uncovered a similar association between a poor father-son relationship and a heightened risk of cancer, and established that it was not merely a by-product of other risk factors such as smoking.

Compelling evidence for links between stressful life events and subsequent breast cancer emerged from a study conducted at King's College Hospital in London, the results of which were published in the *British Medical Journal* at the end of 1995. Women who had a suspicious but (at that point) undiagnosed lump in their breast underwent a psychological interview before a biopsy was carried out. When the results were analysed it transpired that women who had experienced severe life events during the preceding five years had a substantially higher probability of being diagnosed as having breast cancer rather than a benign lump or no disease. Nearly half the women who had experienced major adverse life events turned out to have breast cancer, compared with less than a fifth of those whose biopsy results proved negative.

Cancer has been linked to depression, too. In the course of a long-term study of more than two thousand middle-aged men, researchers at the University of Illinois found that men who showed signs of depression had a significantly greater risk of developing and dying from cancer over the following twenty years. This connection between depression and subsequent cancer held up even when the data were adjusted to take account of other risk factors such as age, smoking and family history of cancer.

Humans are not the only species to suffer from cancer and neither are we the only animals in whom psychological factors have a bearing on cancer. Decades of experimental research with animals (some of it rather grisly, it has to be said) have established beyond reasonable doubt that psychological stress can influence the formation and growth of tumours. Various psychological stressors, including rotation, crowding, handling, isolation, electric shocks, loud noises and physical restraint, have been shown to influence animals' susceptibility to cancers, and the rate at which those cancers develop and spread.

The results of these animal experiments have been diverse and complex. What they show is that although stress usually makes tumours grow faster, it can have the opposite effect under certain circumstances. The way in which a stressor affects tumour growth depends critically on those basic qualities that we considered in chapter 5: intensity, duration, timing, predictability and control.

In the 1970s an American scientist, Vernon Riley, discovered that one way of stressing a mouse without being brutal is to rotate its cage on a modified gramophone turntable (remember those?) for, say, ten minutes in every hour. Rotating the mouse in this way consistently evokes a stress response, including the tell-tale release of glucocorticoid hormones from the adrenal glands. The stress response to rotation is dose-dependent; in other words, the faster the cage is rotated, the higher the occupant's glucocorticoid hormone levels. After an hour or two of this treatment the number of T-lymphocytes circulating in the animal's blood drops. After twenty-four hours there is a measurable reduction in the size and weight of its thymus. More importantly, rotation stress enhances the growth of some types of tumour, especially slow-growing tumours which are attacked by the immune system.

Control has a substantial bearing on whether stressors promote the growth of tumours in animals. You may recall that controllable stressors generally evoke a smaller stress response than uncontrollable (but otherwise similar) stressors. By and large, uncontrollable stressors are bad news for an animal with a tumour. Tumours develop sooner, grow faster, reach a larger size and are more likely to prove fatal in animals subjected to uncontrollable stressors, compared with animals subjected to identical amounts of controllable stressors (or no stressors at all). In fact, moderate controllable stressors often have little or no discernible impact on tumours.

Madelon Visintainer and colleagues at the University of Pennsylvania investigated the effects of control on the ability of rats to reject tumours. The rats were implanted with a small tumour and then given stressful electric shocks. The shocks were either controllable (the rat could escape the shock by pressing a bar) or uncontrollable (pressing the bar made no difference). The ability to control the stressor had a tremendous impact on the rats' ability to reject the tumour: 63 per cent of the rats which had been given controllable shocks successfully rejected the tumour, as against 27 per cent of those which had received identical amounts of uncontrollable shocks. Lack of control over the stressor – a purely psychological variable – more than doubled the rats' chances of succumbing to the tumour.[5]

Clearly, we must be cautious about extrapolating the results of fairly crude animal experiments to humans. Mice and rats are not the same as us. Moreover, electric shocks and other stressors used in laboratory experiments are hardly representative of the real world. That said, basic biological processes are essentially the same in closely related species, so fundamental truths can be gleaned from rats, mice and monkeys.

Another point which bears repetition is that psychological factors probably play an even more prominent role in human disease processes, including cancer, than they do in other species. Our propensity to dwell on past events and fret about the future can vastly extend the period over which stressors are stressful. If psychological stressors affect the growth of tumours in rats, mice and monkeys – which they do – then it is a fairly safe bet that something similar will happen in humans.

Is there a cancer-prone personality?

> What Soft – Cherubic Creatures –
> These Gentlewomen are –
> One would as soon assault a Plush –
> Or violate a Star –
>
> Such Dimity Convictions –
> A Horror so refined
> Of freckled Human Nature –
> Of Deity – ashamed
>
> Emily Dickinson (*c.* 1862)

In recent years scientists have turned the spotlight on a cluster of personality traits thought to be associated with a heightened risk of cancer. This cluster is known as Type C. The Type C pattern shares some of the same conceptual problems as the Type A coronary-prone behaviour pattern, including the lack of a universally agreed definition. There is, nonetheless, a reasonable consensus that certain characteristics are key components of Type C:

- the suppression of strong emotions
- compliance with the wishes of others and a lack of assertiveness
- avoidance of conflict or behaviour that might offend others
- a calm, outwardly rational and unemotional approach to life
- obeying conventional norms of behaviour and maintaining the appearance of 'niceness'
- stoicism and self-sacrifice
- a tendency towards feelings of helplessness or hopelessness

Most researchers agree that the core feature of the Type C personality is the propensity to bottle up strong emotions, especially anger.

SOME FICTIONAL TYPE Cs

The Type C personality is perhaps best illustrated by selecting fictional characters who display the characteristics in an extreme form. The Type C traits of self-sacrificing stoicism, lack of emotional expression and pathological niceness are encapsulated in the mediaeval folk tale of Patient Griselda. Her story has had several renditions, including Boccaccio's *Decameron* and Thomas Dekker's *Patient Grissel*, but her best-known manifestation is the much put-upon Grisildis in 'The Clerk's Tale' from Chaucer's *Canterbury Tales*.

Grisildis is a beautiful and virtuous peasant girl who lives in a village ruled over by a young nobleman called Walter. When Walter's subjects beg him to marry he decides to take Grisildis as his bride. The groom turns out to be anything but a New Man; on their wedding day he tells Grisildis that he requires absolute obedience. She readily agrees and soon wins universal plaudits for her virtue and gentleness. Yet despite her obvious rectitude, Walter decides that he must put Grisildis to the test. He proceeds to do this by subjecting her to a series of pointless and sadistic cruelties over a period of several years.

For an opener, Walter gets one of his officials to confiscate their baby daughter, ostensibly with the intention of murdering the defenceless mite. On being told that this is her husband's command, the obedient Grisildis hands over her daughter without demur. A few years later Grisildis gives birth to a son and Walter does exactly the same thing. Once again, Grisildis surrenders her offspring, proclaiming that above all else she wishes to be subservient to her husband's will.

By this stage, even hard-hearted Walter is starting to be impressed by his wife's impeccable docility. But the brute is still not satisfied. He acquires counterfeit letters from the Vatican which supposedly grant him permission to divorce Grisildis and announces his intention to remarry. Without the tiniest squeak of protest, Grisildis – asking only that he return her peasant dress so that she will not have to walk naked to her village – volunteers to take charge of the wedding arrangements and ensure that the new bride is well looked after. The ghastly Walter at last decides that Grisildis has proved her worthiness. He reveals that it has all been a ruse, returns her son and daughter, and they all live happily ever after.

Throughout her pointless torments Grisildis displays unwavering compliance with her husband's every whim. She bears great suffering in silent passivity, never giving vent to anger or resentment, never indicating the slightest desire to bury an axe in Walter's deserving head, despite immense provocation. The reader sees that inwardly Grisildis experiences great emotional turmoil, but she never expresses it. She is, in short, the perfect human carpet:

> But ful of pacient benyngnytee,
> Discreet and pridelees, ay honurable,
> And to hir housbonde ever meke and stable.

An uncannily accurate portrait of a Type C persona is contained in May Sinclair's novel *The Life and Death of Harriett Frean*.[6] It is a moral fable about the repressed, emotionally starved life of Harriett Frean, a 'perfect daughter of the Victorian age'. Harriett has all the main Type C ingredients. She ruthlessly suppresses her emotions; she is self-sacrificing and does what others expect of her, obeying the wishes of her parents without demur; she rigidly observes the conventional norms of behaviour; she is unemotional; and she is, outwardly at least, frightfully nice. Taught by her genteel Victorian parents to equate love with selflessness, Harriett Frean finds a strange happiness in denying her own desires. Harriett's mother is cast from the same uptight Type C mould. Indeed, she is so excessively self-sacrificing that she refuses an operation to treat her cancer because of the expense and dies as a direct result.

As a young woman Harriett falls in love with Robin, her best friend's fiancé, but then renounces her love with 'a thrill of pleasure in her beautiful behaviour'. When Robin (who is now unhappily married to the best friend) berates her for refusing to marry him, Harriett suppresses her anger like a true Type C and manages to sidestep the almost unavoidable row. In her dry, sexless, middle age, she becomes increasingly isolated from her few remaining friends.

Like her Type C mother before her, Harriett Frean develops cancer and dies. Even when she goes into hospital in the terminal stages of her disease, Harriett behaves impeccably and refuses to open her mouth lest she say anything indecent whilst under anaesthesia.

A male counterpart of Harriett Frean is Henry Earlforward, the miserly Clerkenwell bookseller in Arnold Bennett's *Riceyman Steps*, whom we encountered in chapter 3. Earlforward is a bland, polite, self-sacrificing man of unvarying routine. In middle age he marries a widow, Violet, but their marriage is sterile and passionless. (Violet eventually dies from a fibroid growth in her uterus: another fatal illness laden with psychological symbolism.)

Suppressing any outward sign of emotion is one of Henry Earlforward's Type C hallmarks. Inside, he is a seething mass of anxieties and emotions like anyone else, but the face he presents to the world is one of perfect tranquillity:

> His knee began to ache. His body and his mind were always reacting upon one another. 'Why should my knee ache because I'm bothered?' he thought, and could give no answer. But in secret he was rather proud of these mysterious inconvenient reactions; they gave him distinction in his own eyes. In another environment he would have been known among his acquaintances as 'highly strung' and 'highly nervously organized'. And yet outwardly so calm, so serene, so even-tempered![7]

Earlforward is equally uncomfortable with any display of emotion by others. When Violet flies into an uncharacteristic outburst at Henry's miserliness, he softly leaves the room to avoid conflict. Violet is frustrated by Henry's unexpressive blandness. She declares that she would rather have a husband who knocked her about a bit than live with a 'locked-up, cast-iron safe' like Henry. Despite her own fatal illness, Violet is terrified of disrupting the routine of the Earlforward household. She knows that any departure from the unbending ritual would distress Henry – not that he would give any outward sign of that distress:

> The upset would be terrible if she failed in her daily role;

> Henry would maintain his calm, but beneath the calm 'what a state he would be in!'

Earlforward, like Harriett Frean, dies of cancer.

The theory that Type C attributes like those of Harriett Frean or Henry Earlforward are linked to cancer is more than a fanciful suggestion. There is reasonably sound scientific evidence to support it. Research has implicated various elements of the Type C profile – especially the repression of emotions – in several forms of cancer, including bowel cancer, lung cancer, breast cancer and malignant melanoma.

In one study, Lydia Temoshok and her colleagues discovered that malignant melanoma patients who habitually repressed their emotions had tumours that were thicker, faster-growing and less well infiltrated by lymphocytes: all of which predict a poor outcome. Similarly, when scientists at Johns Hopkins University analysed the incidence of cancer among a sample of 972 doctors they found that those who had been assessed as emotionally suppressed loners when they were students were (astonishingly) sixteen times as likely to develop cancer over the following thirty years as those who were emotionally expressive. Such connections between Type C characteristics and cancer appear to hold true even when other medical risk factors like smoking, diet, age and family history of cancer have been taken into account.

The link between Type C personality attributes and cancer is sufficiently reliable that researchers in several countries have been able to use Type C characteristics to predict which patients will be diagnosed as having cancer after a biopsy. In one study, women who had a suspect lump in their breast were interviewed on the day before a breast biopsy, to see whether they had Type C personality characteristics. On the basis of this psychological assessment alone, the researchers could predict which women's biopsies would show them to have breast cancer (rather than a benign lump) with a success rate of over 80 per cent.

A 1994 American study produced similar results. Scientists at Purdue University assessed the psychological profiles of 826 women who had attended a breast-screening clinic and were awaiting the results of their mammograms. As in the Johns Hopkins study, malignant tumours (as opposed to a benign growth or nothing at all) were most likely to be found in women who exhibited the key Type C trait of emotional repression. In addition, the data established links between a high risk of malignancy and loneliness or recent major life events. Essentially the same conclusions emerged from a British study of two thousand women undergoing a breast examination. Cary Cooper and colleagues at the University of Manchester Institute of Science and Technology found that women who tended to bottle up their feelings – the hallmark of the Type C – were more likely to have breast cancer than benign lumps. The women diagnosed with breast cancer also tended to be the ones who had experienced the most stress within the previous two years.

Another component of the Type C cluster that has often been linked with cancer is a sense of hopelessness. In a study conducted in the 1960s, women undergoing a cervical smear test were interviewed before the outcome of their test was known. None of the women had any overt symptoms of cervical cancer at that stage. The results showed that women who expressed feelings of hopelessness during the interview had the greatest likelihood of being diagnosed as having a malignant tumour. The interview alone predicted the results of the smear test in over three-quarters of cases.

Some remarkable evidence for links between personality and cancer came from research carried out by Ronald Grossarth-Maticek in the former Yugoslavia. In the 1960s Grossarth-Maticek started a long-term prospective study of 1353 Yugoslavian villagers. He discovered that individuals who scored highly on psychological measures of rationality and anti-emotionality (both of which are aspects of the Type C personality cluster) were most at risk from cancer. Moreover, a special type of behaviour therapy that curbed these attributes was found to reduce this risk. Further investigations by Grossarth-Maticek and his collaborator Hans Eysenck uncovered evidence that personality variables and psychological stress have a crucial bearing on the risk of dying from cancer or heart disease years later. Indeed, so strong were the associations reported by Grossarth-Maticek and Eysenck that sceptics dismissed their results as simply too good to be true.

Psychological influences on survival

Do not go gentle into that good night . . .
Rage, rage against the dying of the light.

Dylan Thomas (1952)

I will be conquered, I will not capitulate.

Samuel Johnson (1784) on his illness,
quoted in Boswell's *Life*

We turn now to the question of survival. As well as influencing our chances of developing cancer in the first place, psychological and emotional factors can also modify our chances of survival and recovery from the disease.

Not surprisingly, severe psychological stress tends to make things worse for cancer patients. There is evidence that it can shorten their survival time. For example, research at Guy's Hospital in London found that severe stress heightened the risk of relapse in women who had been operated on for breast cancer. Women who had experienced seriously adverse life events such as job loss or bereavement during the post-operative period were five times more likely to have a recurrence of breast cancer than women who had not experienced severe stress (but whose medical condition and backgrounds were similar). Stress makes matters worse, but can other mental states achieve the opposite? It would appear so.

Improving the state of cancer patients' minds can certainly make them feel better, both emotionally and physically. A 1995 review used meta-analysis to draw together the results of forty-five separate studies which had investigated the benefits of various types of psychological therapies in the treatment of cancer patients. The meta-analysis confirmed that psychological interventions such as relaxation training, cognitive therapy and social support, could bring about marked improvements in physical symptoms and pain control, as well as helping patients to adjust emotionally to their disease. But can psychological therapies help cancer patients to survive?

We are all familiar with the notion of individuals 'fighting' their cancer. This battlefield metaphor has become a media cliché; people no longer die from cancer, they lose their battle with it. Scientists have investigated the notion and found compelling evidence to suggest that social environment and mental attitude can modify our chances of surviving cancer.

Considerable publicity has been generated by the suggestion that cancer victims who display a fighting spirit have a better chance of surviving than those who respond stoically, or those who give up and helplessly accept their plight. This idea is underpinned by some respectable scientific research.

In a trailblazing study, Keith Pettingale, Steven Greer and colleagues at the University of London tracked the progress of a group of women who had been diagnosed with early-stage breast cancer. A few months after the initial diagnosis the researchers assessed each woman's psychological response to her disease. Five years later they checked to see how each woman had fared medically. The results showed that a woman's psychological reaction to her disease had a definite bearing on her likelihood of surviving without any recurrence of cancer over the next five years. Those who responded to the initial diagnosis by displaying either a fighting spirit or denial fared better than those who stoically accepted their condition or responded with feelings of helplessness or hopelessness.

When the researchers conducted a follow-up survey of the same women fifteen years later, those who had originally displayed a fighting spirit or denial continued to fare better: 45 per cent of them were still alive and free from cancer, as against 17 per cent of the women who had reacted initially with stoic acceptance, helplessness or anxious preoccupation. Indeed, a woman's initial psychological response proved a better predictor of her survival than the initial size of her tumour. Moreover, it appeared to act independently of other factors affecting survival, including the initial size of the tumour, the woman's age and the treatments used.

A number of studies have exposed similar connections between patients' psychological responses to their cancer and their subsequent survival. A study of patients who had received a bone marrow transplant for leukaemia found that those whose attitude towards their cancer was characterized by 'anxious preoccupation' (the antithesis of denial or a fighting spirit) had a poorer chance of survival. Results such as these suggest that it may be better for a cancer sufferer to have a strong emotional reaction, of whatever sort, than no reaction at all.[8]

The discovery that cancer patients who react with a fighting spirit may survive longer ties in with psychological research on the Type C personality. Researchers at the Cancer Research Campaign in England have found that Type Cs are inclined to react with feelings of hopelessness or helplessness when told they have cancer. A study of women with early breast cancer revealed that those who displayed the Type C characteristic of controlling their anger and other emotions also tended to have a fatalistic, helpless attitude towards their cancer.

The patient's own mental state is not the only thing that matters, of course. Other people are important, too. In view of all we have seen about the influence of social relationships on health it should come as no surprise that the quantity and quality of a person's relationships can have a bearing on their chances of surviving cancer. There is growing evidence that cancer patients who have good personal relationships and social support tend to survive longer than those who are lonely or socially isolated. A Canadian study of patients with newly diagnosed lung cancer revealed that the odds of the patient dying within a year were greater in cases where social support was lacking. Studies of women with breast cancer have confirmed this: women who receive lots of emotional support from their partner, family, friends and doctor tend to survive significantly longer.

Perhaps the most striking evidence for the benefits of social relationships comes from the work of David Spiegel and his colleagues at Stanford University. To their surprise, they found that women with advanced breast cancer survived twice as long when they took part in psychological therapy which improved their social environment.

The therapy used by Spiegel's team consisted of weekly ninety-minute sessions in which the patients and medical staff met in a group to discuss the patients' feelings and thoughts about their cancer. The women were encouraged to confront their fears of dying and vent their emotions, in a setting where they had the emotional support of fellow-sufferers and medical staff. They were also taught a simple relaxation technique to help them cope with their pain. One consequence of the group therapy was that the patients soon developed close, mutually supportive relationships with each other.

At the start of Spiegel's study, fifty women with metastatic breast cancer were randomly assigned to receive weekly sessions of the group therapy, in addition to the standard medical treatment. (A further thirty-six women were randomly assigned to a comparison group and received only the standard medical treatment.) The initial results showed that the therapy was highly effective in improving the women's quality of life. Their mood and mental wellbeing improved over time, despite their worsening physical condition, while the mental state of patients in the comparison group deteriorated. The therapy reduced the women's perception of pain as well.

As originally conceived, the therapy had been intended to help cancer patients and their families cope with the disease and augment their quality of life for whatever time they had left. The researchers had not envisaged that the therapy might help patients to live longer, nor were the patients ever led to believe this. At around this time, various New Age gurus were attracting considerable media attention with their assertions that cancer could be cured by mental forces alone. Spiegel and his scientific colleagues were rightly sceptical of the wilder claims. Nevertheless, they decided to see whether their research data could cast any light on the issue. They therefore checked whether the group therapy had in any way affected the patients' survival. To their surprise, it had.

By this stage, all but three of the original eighty-six women had died. The group therapy had clearly not conjured up any miraculous cures. But it had still achieved something pretty remarkable: it had doubled the average length of time the women survived before eventually succumbing to their cancer. Among the comparison group, who received only the standard medical treatment, the average interval between entering the study and dying was nineteen months; for the women who had received group therapy it was thirty-seven months. The average interval between the initial appearance of secondary tumours (metastasis) and death was also significantly longer among those who had received group therapy. When the Stanford team's remarkable findings were published in that august medical journal the *Lancet* in 1989 they generated quite a stir.

Spiegel's discovery and others like it have occasionally been misrepresented by the media and New Age gurus as evidence that people can cure themselves of cancer by purely psychological means. Not so. David Spiegel and other serious scientists working in this field have been careful to avoid making such unfounded claims. Their interpretation (which is almost certainly the correct one) is that a high quality social environment can help to improve a cancer patient's mental and physical state and hence prolong their survival.

Further confirmation that psychological treatments can help to extend the survival time of cancer patients has come from the work of Fawzy Fawzy and colleagues at the University of California at Los Angeles. They looked at the effects of psychological therapy on the mental and physical health of patients with malignant melanoma. Each patient received a ninety-minute session of therapy once a week for six weeks, starting shortly after the diagnosis and initial treatment. The therapy was designed to educate the patients about their health, reduce their fear, improve their coping skills, teach them how to manage stress and bolster their social support. A comparison group of patients received no psychological therapy but were in other respects similar.

As in the Stanford research, this modest psychological intervention turned out to have a greater impact on patients' survival than had been anticipated, as well as improving their mental wellbeing and quality of life. Over the next five or six years, malignant melanoma patients who had received the psychological therapy were only half as likely to suffer a recurrence of cancer as patients in the comparison group. The therapy also reduced the mortality rate by a third. Other researchers have found that psychological and social treatments can appreciably lengthen the survival time of patients with leukaemia or cancer of the lymph nodes (lymphoma).

Fawzy's research also cast light on the biological mechanisms underlying these remarkable phenomena. The cancer patients who received the psychological therapy exhibited changes in their immune function. Compared with patients who received no psychological therapy, they had significantly higher numbers of circulating lymphocytes. In particular they had a greater quantity of natural killer cells and large granular lymphocytes, both of which can kill tumour cells. Their natural killer cells were more active, too.

One interpretation of this finding is that the psychological therapy somehow stimulated the patients' immune systems and thereby enhanced their bodies' ability to resist the further growth of their tumours. It is time to investigate the question of mechanism.

How does it work?

By what means could our mental state alter our susceptibility to cancer or our chances of surviving it? To find the answer we must return to the ubiquitous trinity of perception, behaviour and immunity.

Behaviour is an obvious mechanism by which the mind can influence our chances of being stricken with cancer. Various behaviour patterns heighten the risk: smoking, excessive sun-bathing and eating a low-fibre diet are well-documented examples.

Perception plays a more subtle and pervasive role. A person's psychological state, emotions and attitudes can affect the speed with which their cancer is detected and even the type of medical treatment they receive.

The chances of surviving many cancers are considerably improved if the disease is detected at an early stage. People who notice and respond to suspicious symptoms have a better chance of spotting their cancer at an early and potentially treatable stage. The same applies to those who regularly examine themselves for suspicious lumps or unusual moles, and those who attend routine screenings for breast or cervical cancer. Conversely, those who ignore the early warning signs, or fail to make use of routine health screenings, run the risk of not receiving expert medical attention until it is too late. It is estimated that routine screening for breast cancer reduces the mortality from this disease by a quarter, so individuals who, for whatever reason, forego the opportunity are exposing themselves to avoidable risks.

The reasons why some potential cancer victims behave prudently while others gamble with their health are complex and poorly understood. The cost and availability of medical care are both factors, but a person's behaviour in this sphere will not be governed solely by rational decision-making or a conscious balancing of costs and benefits. Emotions play a crucial role.

Psychologists have found that women who voluntarily attend screenings for breast cancer have emotional characteristics distinct from those of women from comparable backgrounds who opt not to be screened. In particular, women who attend screenings typically report that they experience more unpleasant emotions than non-attenders. Perhaps they are impelled by their anxieties and unhappiness to attend health screenings. Whatever the psychological and emotional antecedents, their cautious behaviour will lessen the statistical risk of their dying from breast cancer.

Perceptions, attitudes and emotions have a major bearing on how quickly people seek medical care when they discover a potential warning sign such as a lump or an odd-looking mole. Some individuals prevaricate, hoping the symptom will disappear or turn out to be trivial. Some will block it out of their minds altogether. Others will immediately become anxious and rush to their doctor for advice.

Complacency, ignorance and fear of embarrassment prevent many men from seeking timely medical attention for the early signs of testicular or prostate cancer. Because of this reticence, a number of men die unnecessarily from what might have been curable cancers. Cancer of the prostate now kills nine thousand men a year in Britain alone and is the third most common cancer among men, although it still receives far less public attention than breast and cervical cancers.

Once a person has started to receive medical treatment for cancer, the final outcome can depend on how well they adhere to the treatment regime. Radiotherapy is a primary tool in the treatment of many cancers, and strict adherence to the radiotherapy schedule has been shown to improve its effectiveness. Much the same is true of chemotherapy. A patient's mental state – particularly their mood and general attitude towards their disease – can affect their compliance with programmes of chemotherapy or radiotherapy and, hence, their survival. For instance, researchers at the Veterans Affairs Medical Center in Memphis found that women with primary breast cancer who displayed high levels of fighting spirit or anxiety were more likely to comply with their chemotherapy regime than women who felt guilty about their breast cancer. You will recall that a fighting spirit is associated with longer survival. Perhaps one way in which it helps is by making cancer patients determined to get the very best out of the medical treatment they are receiving.

Cancer patients' mental states will also have an effect on how well they look after themselves. There have been suggestions that eating the right food, taking regular exercise and getting enough sleep can help to slow the progression of cancer. The true effectiveness of these measures remains controversial, although they cannot do any harm and may improve the patient's quality of life. However, if looking after yourself in these ways does make a real difference to the progression of cancer, then having positive attitudes and sufficient motivation could be vital.

Another factor that can be influential in the treatment and outcome of cancer (although in principle there is no reason why it should) is the patient's relationship with his or her doctor. Radiotherapy and chemotherapy can be extremely unpleasant and demanding. In borderline cases a patient's willingness, determination and emotional capacity to withstand such an ordeal will inevitably influence the doctor's decision on whether to give the most vigorous and unpleasant treatment, as opposed to something gentler but less effective.

We come now to the third constituent of our trio: the body's biological defence mechanisms. Innumerable biological mechanisms are involved in a disease process as complex as cancer. Psychological factors therefore have the potential, at least in theory, to act upon cancer in a variety of ways and at different stages in the disease. They might influence the initiation of a tumour, its subsequent growth, or its metastatic spread to form secondary tumours in other parts of the body.

Both tumour growth and metastasis are, to a degree, dependent upon the local blood supply to the tumour, so anything that affects blood vessels might, in theory, have a bearing on tumour growth and metastasis. We have already seen how the mind, via the sympathetic nervous system, can affect blood vessels.

Hormones, too, can have an impact. Prolactin, for example, is known to facilitate the growth of breast tumours, and its secretion is stimulated by severe stress. As are glucocorticoid hormones, which, as we saw in chapter 5, assist the mobilization of the body's energy reserves and augment the level of glucose in the bloodstream. Rapidly growing tumours need lots of energy and are good at absorbing glucose from the blood before other tissues can get hold of it. It has therefore been suggested that stress may encourage the growth of tumours by boosting their glucose supply. Tumour growth might be further encouraged by another action of glucocorticoid hormones, which is to stimulate the growth of capillary blood vessels (a process known as angiogenesis).

In view of this possible link between the hormonal response to stress and tumour growth, it may be noteworthy that Type C individuals who do not express their emotions have been found to show consistently larger stress responses than more expressive individuals under comparable circumstances.

But what, you might be thinking, about the immune system? What do psychoneuroimmunologists have to say about the connections between the mind, immunity and cancer? As you would expect, scientists have been busy exploring the possibility that psychological and emotional factors might modify the initiation or progression of cancer via their effects on immunity. There is little doubt that the mind can influence the immune system. But does the immune system influence cancer?

An appealing but controversial idea about immunity and cancer is that of immune surveillance. This is the idea that cancer cells spontaneously arise within the body, but are normally detected and eliminated by the immune system before a tumour can develop. If this were true, then anything that impaired the immune system's ability to carry out its surveillance function would heighten the risk of cancer.

In its crudest form, the theory of immune surveillance is not borne out by the facts. Were it that simple, a person or animal with a suppressed immune system would spontaneously develop a whole range of cancers. In reality, immune-suppressed individuals do not necessarily develop any cancer at all, and those cancers that do appear tend to be restricted to a few specific types. Those whose immune systems are severely weakened by AIDS or immune-suppressive drugs are more vulnerable to non-Hodgkin's lymphoma and certain skin cancers. The best known of these is Kaposi's sarcoma, a once exotic skin cancer which has become a common affliction of AIDS victims. With these few exceptions, however, immune suppression does not lead to an indiscriminate outbreak of cancers.

It has become clear that the immune system plays a much greater part in fighting certain forms of cancer than others. The body's immune response has a significant impact on malignant melanoma and renal cell carcinoma, for example, though other cancer cells may trigger only a weak immune response, if any.

Where the immune system does respond, it offers a dual threat to cancer cells. Should the specific response fail to recognize cancer cells as antigens, a non-specific attack can still be mounted using natural killer cells and macrophages. Natural killer cells are believed to have an important role in preventing metastasis, the spread of malignant tumours to other parts of the body. Dana Bovbjerg and colleagues at the Memorial Sloan-Kettering Cancer Center in New York discovered that natural killer cell activity was consistently lower in healthy people who had a family history of cancer (and who were therefore at greater risk of developing the disease themselves).

More direct evidence for the involvement of natural killer cells came from research carried out at the Pittsburgh Cancer Institute. This found that breast cancer patients who perceived themselves as lacking social support had lower levels of natural killer activity and an increased risk of secondary tumours spreading to their lymph nodes. Here we see a convergence of all three elements in the psychoneuroimmunological story: an adverse psychological influence (lack of social support), reduced immune function and an increased rate of tumour metastasis.

Links between psychological factors, natural killer cell activity and cancer have been uncovered in other species as well. Scientists at the University of California at Los Angeles found that exposing rats with mammary tumours to acute psychological stress substantially reduced their natural killer cell activity and doubled the metastatic spread of secondary tumours to their lungs. This lends support to the theory that stress can encourage the metastatic spread of tumours by impairing the activity of natural killer cells.

A series of intriguing experiments have gone one step further and unearthed a link between heritable personality traits, natural killer cell activity and susceptibility to cancer in mice. (Yes, mice do have distinctive personalities.) J. M. Petitto and colleagues at the University of Florida investigated immune function in a strain of mice that had been selectively bred to behave in an unaggressive and socially inhibited way. When confronted with an unfamiliar mouse, these 'shy' mice exhibited none of the species-typical aggressive behaviour: they preferred to avoid a fight. (Shades of Type C, you may be thinking.) The researchers also discovered that the shy mice had a greater vulnerability to cancer and measurably lower levels of natural killer cell activity. Obviously it is a huge jump from an inbred laboratory mouse to Harriett Frean – but the parallels are tantalizing.

Psychological factors like stress might, in addition, be able to impair the body's defences against cancer by altering the molecular mechanisms responsible for repairing faulty DNA in cells. Janice Kiecolt-Glaser, Ronald Glaser and their colleagues at Ohio State University have discovered that DNA repair is less effective in the lymphocytes of humans (or rats) suffering from high levels of stress. Psychological stress reduces the ability of lymphocytes to manufacture an important enzyme used in the DNA repair process. This discovery is potentially far-reaching because a person with faulty DNA repair mechanisms has a greater risk of getting cancer. The majority of cancer-causing agents work by damaging the DNA in cells. Psychological stress which impairs the efficiency with which cells correct errors in their DNA can, in principle, increase the risk that those cells will become cancerous.

An honest debate about the role of psychological variables in cancer has undoubtedly been impeded by two things, over and above the usual scepticism which attaches to anything tarred with the psychosomatic brush. First, there is the emotionally charged nature of the disease itself. The topic of cancer remains an upsetting one. Then there are the overblown, unsustainable claims of mental miracle cures emanating from the legions of New Age gurus who bedevil this field. We shall return to the subject of miracle cures later. One of their many undesirable consequences is to raise the barriers of medical scepticism even higher, further impeding a true understanding of the disease.

There is now little doubt that psychological and emotional factors can influence the development and progression of certain cancers, although the practical importance of their effects remains to be explored. Psychological variables such as social support and the patient's emotional response to their disease have a substantial bearing on survival. Equally, there is little doubt that appropriate psychological and social interventions can be of enormous benefit in helping cancer patients to cope with their disease, improving the quality of what remains of their lives and, in some cases, extending their survival time. To ignore the mind is an irresponsible waste.

10

Encumbered with Remedies

Suit the action to the word.

William Shakespeare, *Hamlet* (1601)

In 1552 Hugh Latimer, the English Protestant martyr-to-be, observed that:

> A great many of us, when we be in trouble, or sickness, or lose anything, we run hither and thither to witches, or sorcerers, whom we call wise men . . . seeking aid and comfort at their hands.

How true this still is. The time has come for us to consider how the interactions between mind and body can be exploited for the practical benefit of humanity. Scientists have uncovered some profound truths about human nature whilst exploring the connections between mind, immunity and disease. Yet we live in a material world where the value of discoveries is measured not by their scientific interest or cultural significance but by their practical utility. So, can this scientific knowledge about mind and body be used to cure people or stop them from becoming ill in the first place? Does it really work?

We have seen evidence that psychological interventions can be of real benefit in heart disease and cancer. In this chapter we shall take a critical look at an assortment of the therapeutic applications of mind–body interactions and the mythology that surrounds the whole subject.

Having discovered that our minds can damage our health, researchers have naturally searched for ways of improving health instead. Amongst other things, they have probed the curative powers of hypnosis, relaxation, physical exercise and numerous forms of behavioural therapy. So far the results have been moderately encouraging, although they fall far short of what certain zealots have claimed.

Relax!

Rest, rest, perturbèd spirit.
William Shakespeare, *Hamlet* (1601)

Since the dawn of time people have been trying to escape the pressures of life by curling up somewhere (metaphorically or literally) and relaxing. Doing nothing is the simplest approach. Those in search of more profound states of relaxation have turned to all manner of structured methods including prayer, yoga and meditation.

Aficionados of the New Age arts have refined these traditional approaches several stages further. Relaxation is no longer simply a matter of relaxing. Those in pursuit of repose must now choose between copious varieties of relaxation techniques, each one a distinct skill that must be learned with devotion (and preferably at great expense). The serious enthusiast uses biofeedback to induce a state of profound physiological rest. But mockery would be misplaced, for science has begun to confirm the essential validity of a common experience. It seems that relaxation can indeed have beneficial effects on the body as well as the mind.

Appropriate relaxation techniques can relieve feelings of anxiety or depression and create a sense of psychological wellbeing. Used in the right way, relaxation assists those who are experiencing high levels of psychological distress or pain; patients undergoing unpleasant medical treatments, for example, or women during childbirth. Controlled scientific studies have established that relaxation training can help to alleviate depression, fatigue, anxiety and unpleasant emotions among cancer outpatients receiving radiotherapy. Healthy individuals who practise relaxation techniques usually experience improvements in their enjoyment of life, personal relationships and performance at work. Properly done, relaxation is a definite aid to mental wellbeing.

But relaxation is not merely a palliative for the mind. It affects the whole body. When a person is deeply relaxed their blood pressure, pulse rate and breathing rate fall, their muscles relax and their coronary arteries dilate – the opposite, in fact, to what happens during a stress response. The regular practice of relaxation techniques can bring about long-term reductions in resting blood pressure and (probably) in the risk of coronary heart disease. For these reasons, relaxation is often a useful adjunct to drugs in the treatment of patients with borderline hypertension.

Many relaxation techniques appear to exploit the phenomenon of conditioning – the associative learning process that we encountered earlier. During the training period the individual unconsciously learns to associate a previously neutral stimulus, such as a particular sound or thought or smell, with the sensation of mental and physical relaxation. By repeatedly pairing the neutral stimulus with relaxation, the stimulus is eventually capable of eliciting the relaxed state by itself. All the individual need do is croon the mantra, imagine the scene, smell the aroma, take the deep breath or hear the piece of music, and the conditioned response of mental and muscular relaxation automatically follows.[1]

There is mounting evidence that relaxation can go further than merely lowering blood pressure. Janice Kiecolt-Glaser and her psychoneuroimmunological group at Ohio State University have demonstrated that relaxation can have beneficial effects on immune function in groups as diverse as healthy young medical students and geriatrics living in retirement homes. Medical students who practised hypnotic relaxation techniques during the month before an important and stressful exam had a higher proportion of helper T-lymphocytes in their blood than students who did not relax. The relaxers also experienced less psychological distress. Relaxation was of benefit to the elderly, too. A month-long programme of relaxation training produced a marked increase in their natural killer cell activity and lymphocyte responsiveness, together with improvements in their immune control over latent herpes viruses and psychological wellbeing. Other experiments have established that relaxation can temporarily augment the levels of antibodies in the blood and saliva, although the clinical significance of these effects is at times unclear.

Relaxation therapy in various guises has proved to be helpful in the treatment of medical problems as diverse as mild hypertension and recurrent mouth ulcers. It has been used to treat alopecia universalis (abnormal hair loss), a condition in which psychological stress and immunological defects play a rather obscure role and which does not respond well to drugs. Japanese investigators found that relaxation was beneficial in five out of six cases of alopecia universalis and was accompanied by changes in T-lymphocytes and beta-endorphin levels. American research has shown that progressive muscle relaxation can reduce the recurrence of genital herpes in some sufferers.

All in all, the case for relaxation is strong. At the very least, it makes people feel better and has no harmful side effects. At best, it can bring significant improvements in immune function and physical health.

Finally, it should go without saying (but won't) that relaxation need not involve physical inactivity. Engaging oneself in an absorbing and reasonably pleasant pastime can also be a healthful tonic to the system. That is precisely why three men (to say nothing of the dog) decided to go on the boating expedition up the River Thames that is so memorably recounted in Jerome K. Jerome's *Three Men in a Boat:*

> . . . we refilled our glasses, lit our pipes, and resumed the discussion upon our state of health. What it was that was actually the matter with us, we none of us could be sure of; but the unanimous opinion was that it – whatever it was – had been brought on by overwork.
>
> 'What we want is rest,' said Harris.
>
> 'Rest and a complete change,' said George. 'The overstrain upon our brains has produced a general depression throughout the system. Change of scene, and absence of the necessity for thought, will restore the mental equilibrium.' . . .
>
> . . . George said:
>
> 'Let's go up the river.'
>
> He said we should have fresh air, exercise and quiet; the constant change of scene would occupy our minds (including

what there was of Harris's); and the hard work would give us a good appetite, and make us sleep well.

Exercise!

The wise, for cure, on exercise depend.
John Dryden, *Epistles* (1700)

Exercise is bunk.
Henry Ford (attrib.)

Just as the symptoms of so-called psychosomatic illnesses have a habit of reflecting the prevailing cultural norms, so too do the methods we employ in our attempts to stay healthy. Nowadays, this means a preoccupation with the very antithesis of relaxation – namely, physical exercise.

Numerous claims have been made for the health-giving properties of physical exercise, some of them wildly over-stated. Nonetheless, the scientific evidence implies that even the most sceptical couch potato would be wrong to dismiss the benefits of exercise. (Being a trifle podgy and indolent myself I would prefer not to dwell on this, but the truth must out.)

Physical exercise, like relaxation, affects both mind and body. Repeated exercise produces a physiological training effect, in which the work capacity of the heart, circulatory system and muscles increases in response to repeated demands. The beneficial consequences of the resulting physical fitness include a reduction in blood pressure, a lower resting pulse rate and less body fat. The statistics show that individuals who exercise regularly have a lower risk of dying from heart disease than those with sedentary lifestyles.

Conversely, lack of exercise is associated with a greater risk of heart disease and other serious disorders, including diabetes mellitus. Lack of exercise is not good for mental health either. A person who takes little or no exercise has a greater statistical risk of suffering depression (although one might need to ponder on which came first: the sloth or the gloom).

I-Min Lee, Chung-cheng Hsieh and Ralph S. Paffenbarger of Harvard University amassed compelling scientific evidence for the long-term health benefits of exercise, which they published in the *Journal of the American Medical Association* in 1995. They examined mortality patterns among 17,321 middle-aged men who had graduated from Harvard University in the 1960s. The data revealed a clear association between physical activity and life span. The more energy a man expended each week in vigorous physical activity, the less likely he was to die prematurely from any cause. A similar link between physical activity and longevity emerged from a long-term study of 8,463 Israeli men. Those who were physically active during their leisure time had a lower risk of dying from coronary heart disease or any other cause over the following twenty-one years.[2]

There is even evidence that regular, moderate exercise may slightly reduce the risk of certain forms of cancer, especially cancers of the colon, breast and uterus. A nineteen-year study of over a million Swedish men found that those who led sedentary lives had a 30 per cent greater risk of developing cancer of the colon.

How might physical exercise protect us against disease? Besides its well-known effects on the capacity of muscles, heart and lungs, what else does regular exercise do to the body? Enter the immune system once again.

Vigorous physical activity has wide-ranging effects on the immune system, not all of which are benign. One well-known immunological result of vigorous exercise is a temporary increase in the number of white blood cells circulating in the bloodstream. This phenomenon, which is known as leucocytosis, was discovered at the turn of the century. Blood samples taken from four runners immediately after they completed the 1901 Boston Marathon showed that their white blood cell counts were between three and five times higher than normal. Exercise-induced leucocytosis is, however, only a transient phenomenon. White cell counts usually return to normal within hours, and regular exercise does not substantially alter the number of white cells in circulation when the individual is resting. For most of the time an athlete's white blood cell count is no higher than that of the average couch potato.

The medical significance of this short-term rise in the white blood cell count is unclear. It may not make much difference to health. However, exercise does have other, potentially more interesting, effects on the immune system. Moderate exercise can elicit temporary increases in several measures of immune function, including the phagocytic (cell-eating) activity of certain white blood cells; the level of interleukin-1 (a cytokine which stimulates other elements of the immune system); the number and activity of natural killer cells; and, possibly, the levels of interferons (another important category of cytokines).

There are indications that the effects of moderate exercise on the immune system are more pronounced in the elderly than the young. This may be because the baseline level of immune function is generally lower in the elderly, so a gentle nudge makes a bigger difference. However, all of these immunological effects of exercise are short-lived, and it is again unclear whether they have any tangible impact on physical health.

But even if exercise brought no physical benefits whatsoever, the psychological benefits alone would make it worthwhile. Regular physical exercise can markedly improve our mood and self-esteem, with beneficial spin-offs in other areas of life such as social and sexual relationships, and performance at work. It can alleviate mild depression: people who exercise regularly and maintain a high level of physical fitness tend to experience less anxiety and depression; they feel more positive and in control (and remember the importance of control), sleep better and enjoy greater self-esteem. Improvements in physical appearance, sexual attractiveness and self-image may add a few extra cherries to the cake.

Another credo of those who preach the gospel of exercise is that it helps to cope with stress. It is widely believed that physical fitness makes us mentally robust and better able to cope with the slings and arrows of daily life. The scientific evidence to support this belief is suggestive, if not entirely conclusive. In one study, researchers assessed the amount of psychological stress that student volunteers had experienced over the preceding year, together with their current physical fitness. The volunteers' health was then monitored over the following months. Individuals who had experienced lots of stress during the preceding year and who were unfit proved to be vulnerable to illness, but stress had relatively little impact on the health of those who were physically fit. A parallel link emerged between stress, fitness and psychological wellbeing. The unfit, stressed students were more depressed than those who were equally stressed but physically fit.

Of greater interest is the suggestion that exercise can reduce an organism's biological reactivity to stressors. Some research has indicated that regular aerobic exercise can decrease the magnitude and duration of the cardiovascular and hormonal reactions to psychological stressors. If true, this could be highly significant. As we have seen, reactivity to stressors is of potential relevance to coronary heart disease and other serious disorders. Yet the evidence on this point is by no means consistent: several well-conducted studies have failed to find any moderating effect of fitness on people's stress response.

For those of us who prefer to take life at a sluggardly pace and eschew the sweaty jogging shoe, the picture is not one of unremitting gloom and doom. People who take up exercising and then abandon it may actually be worse off than those of us who never even started. Ronald Grossarth-Maticek, Hans Eysenck and colleagues at the University of London found that men who were once involved in regular sporting activity, but had given it up, had a higher mortality rate than those who had never played sports at all. (Needless to say, men who continued to be actively involved in sport had the lowest mortality rate of all.) And then there is the not insignificant matter of sports injuries.

Like most remedies, physical exercise is a two-edged sword. In moderation it is almost entirely a Good Thing, both for physical and mental health. In excess, however, it can do harm. Extreme physical exertion is a powerful stressor which can impair immune function and lower the body's defences against infection.

At full fling, the body's rate of energy consumption increases to ten times its resting level, while in certain muscles it can increase by a factor of fifty. To satisfy this rise in demand the heart must beat three times faster than normal and must quadruple the rate at which it pumps blood, to around 20 litres per minute. (It achieves this by pumping more blood with each beat, as well as beating faster.) This huge increase in the cardiovascular workload can be catastrophic for a person whose heart or coronary arteries are weakened by disease. Violent exercise can aggravate other disorders as well. For example, the increase in respiration that accompanies physical exertion can trigger asthmatic attacks in susceptible individuals.

In many respects the immediate biological effects of intense exercise are akin to those induced by stressors. Extreme physical exertion elicits similar hormonal changes to those that occur during the stress response, including the release of adrenaline, noradrenaline, cortisol and endorphins.

Intense exercise also produces some undesirable changes in immune function. These changes include a drop in the secretion of two main classes of antibodies (IgA and IgM), a decline in the number and responsiveness of circulating lymphocytes and a drop in natural killer cell activity. Moreover, although exercise temporarily boosts the total number of white blood cells circulating in the bloodstream, it does not affect all white blood cells uniformly. The number of circulating suppressor/cytotoxic T-lymphocytes grows out of proportion to helper T-lymphocytes – a shift that could have undesirable repercussions for immune function.

There is convincing evidence that keep-fit fanatics who engage in excessive physical exercise are more susceptible to coughs, colds and other infections of the respiratory tract, probably as a result of the hormonal and immunological changes that accompany extreme exertion. Top athletes are notoriously more prone to minor infections than ordinary mortals. They might be extremely fit, but it does not follow that they will be extremely healthy. Too much exercise can also produce undesirable changes in mental state. Experimental studies with volunteers have found that exercising to the point of exhaustion produces mood disturbances, difficulties in sleeping and loss of appetite, as well as impairments in immune function.

Those little pink pills

> For it is with the mysteries of our religion, as with wholesome pills for the sick, which swallowed whole, have the virtue to cure; but chewed, are for the most part cast up again without effect.
>
> Thomas Hobbes, *Leviathan* (1651)

Among the most potent, tried and trusted remedies that rely entirely on psychology is the placebo effect – that remarkably robust phenomenon whereby we get better because we *believe* we are going to get better.

People who take inert placebo 'drugs', believing them to be real drugs, typically experience a marked reduction in pain or other symptoms even though the placebo is chemically inactive. The placebo effect is remarkably powerful, producing noticeable improvements in at least a third of all patients, often more. Placebo drugs and placebo medical procedures have proved to be effective against a wide range of medical problems including chronic pain, high blood pressure, angina, depression, schizophrenia and even cancer. They are the twentieth century's continuation of an ancient tradition of secret potions, incantations, leeches, the royal touch, pilgrimages to healing shrines, gold tablets and magic spells.

The placebo effect is a palpable demonstration of how our psychological expectations can override the signals coming from our body. Placebos only work well if the patient believes in them. The effectiveness of a placebo drug or procedure is greatly improved if the doctor convinces the patient that the treatment will make them better, and if the placebo is administered in a way which increases its psychological potency. For instance, placebos generally work better when injected rather than taken as tablets. An injection has a bigger psychological impact than swallowing a pill. Pharmaceutical companies have discovered that when a placebo is taken in tablet form, the colour, size and shape of the tablet all have a bearing on its effectiveness. Those little pink pills may have to be pink rather than white. As psychologist

Richard Totman so aptly put it, 'take anything that is either nasty, expensive or difficult to obtain, wrap it up in mystery, and you have a cure.'

The psychological and physiological mechanisms underlying the placebo effect are complex and not well understood. Some scientists have suggested that it is mediated by endorphins, the pain-relieving opioid chemicals released by the brain during stress. Others have suggested that it may rely on some form of conditioning. Whatever the mechanisms of the placebo effect, there is no doubt that the mind lies at its very centre.

Psychoneuroimmunology and AIDS

> When a lot of remedies are suggested for a disease, that means it can't be cured.
>
> Anton Chekhov, *The Cherry Orchard* (1904)

In recent years much of the effort to reap the rewards of psychoneuroimmunology has been focused on the problem of AIDS. This makes sense. As well as being one of the most pressing challenges facing medical science, AIDS is quintessentially a disease of the immune system.

The human immunodeficiency virus (HIV), the causative agent of AIDS, wreaks most of its havoc by destroying the victim's helper/inducer T-lymphocytes. Given the myriad connections between mind and immunity, it follows that psychological and emotional factors might, in theory at least, play a part in the development and progression of AIDS. So far, however, the outcomes of the research have failed to live up to the high expectations.

Several years can elapse between a person's initial infection with HIV and the appearance of AIDS. When someone is first infected with the HIV retrovirus[3] they may experience transient, 'flu-like symptoms. The virus then enters a period of apparent dormancy when no symptoms are noticeable. (Whether the viruses are actually dormant is another matter; recent evidence shows they are not.) Perhaps years later the virus begins to cause serious damage to the immune system and the symptoms of AIDS appear. Some people develop AIDS within two or three years of infection while others have remained free from symptoms for over a decade. And when AIDS does develop, its rate of progression can vary enormously, with some individuals succumbing rapidly while others are still alive after five years.

The immune system plays a critical role in suppressing the HIV retrovirus during the latent period between initial infection and the onset of full-blown AIDS. In theory, an agent that further weakened the immune system of an HIV-infected person – psychological stress, for example – might also hasten the onset of AIDS. Conversely, if we wish to delay the onset of AIDS or slow its rate of progression then we should look for ways of enhancing the immune function of those infected with the virus. Psychoneuroimmunologists have been feverishly exploring both possibilities, but so far with limited success.

There is a growing body of evidence that psychological factors can influence the immune function of those infected with HIV, although the implications for their health remain unclear. Karl Goodkin and colleagues at the University of Miami discovered links between immune function and stress in a study of gay men who were infected with HIV but had not yet exhibited any symptoms of AIDS. The men's natural killer cell activity and lymphocyte counts were related to the amount of psychological stress they had experienced and their ability to cope with that stress. The men with the best immunological indicators were the ones who had experienced the least stress and had dealt positively with it.

Scientists have also been investigating the potential benefits of psychological treatments aimed at boosting immune function in HIV-infected individuals, such as stress management, relaxation techniques and behavioural therapies. So far, the results have been moderately encouraging but inconclusive. In one study, for example, a programme of twice-weekly sessions of muscle relaxation, meditation and hypnosis led to significant improvements in a group of asymptomatic HIV-positive men. As well as improving their mood and mitigating their anxiety, this therapy boosted the men's

T-lymphocyte counts. Yet it remains debatable whether psychological factors make a tangible difference to the immune function and physical health of HIV-infected people. Their mental health is another matter, of course, and there is no question that appropriate psychological assistance can improve the quality of life of AIDS victims.

The potential benefits of physical exercise are also being explored. As we saw earlier, moderate exercise can augment the number of circulating T-lymphocytes and have a range of positive effects on immune function. In theory, such changes might help to slow the progression of AIDS, but this remains largely speculative. A suitable programme of moderate physical exercise improves the psychological wellbeing of those infected with HIV, much as it does for anyone else. By relieving feelings of anxiety and depression, exercise can mitigate the emotional pressures that inevitably result from having a fatal condition. However, there is as yet little solid evidence that exercise alone can enhance the immune function and physical health of those who are HIV-positive or have AIDS.

Research conducted at the University of Miami did achieve promising results. Unsurprisingly, most people become upset and depressed when told of their HIV-positive status, and this emotional reaction is often accompanied by changes in immune function, notably a drop in natural killer cell activity. The Miami study found that individuals who were physically fit and engaged in a programme of aerobic exercise underwent less severe emotional and immunological reactions to the bad news.

The immune system is not the only part of the victim to be touched by HIV and AIDS. The mind feels the effects, too. For a start, the HIV retrovirus can physically damage the central nervous system and cause a variety of psychiatric disorders, as we saw in chapter 4. However, the psychological and emotional reverberations of HIV infection start long before this. A person who knows they are infected with HIV, or fears that they might be, will be highly vulnerable to chronic stress, anxiety and depression. HIV-positive individuals, members of high-risk groups and AIDS victims often have to shoulder the additional burdens of stigma, social isolation and personal loss. Statistically, they have a higher than average risk of losing a partner or close friend to the disease. We have seen abundant evidence that both social isolation and bereavement spell bad news for health in their own right; mixed together, these psychological ingredients make a dangerous cocktail.

Sanford Cohen of the Boston University School of Medicine has drawn a thought-provoking analogy between the enormous psychological pressures that accompany AIDS and the phenomenon of voodoo death. Both the AIDS victim and the victim of voodoo death are informed by authority figures that they will inevitably die. Both accept this outcome as certain and unavoidable. Both experience feelings of hopelessness, helplessness and isolation. These feelings are often reinforced by the responses of the victim's family and friends, who turn away from the victim and behave as though he or she is no longer a normal human being.

In this sense, Cohen has argued, being told that you have AIDS is metaphorically akin to having the voodoo bone pointed at you. The psychological and emotional consequences of the two situations might have something in common. It is conceivable that both could harm the victim's immune system and health in similar ways. A combination of severe stress, hopelessness and social isolation might further impair the AIDS victim's immune function and hasten their demise. It is certainly food for thought.

One point on which most scientists in this field do agree is that intervention must occur at the earliest possible stage in the disease process. Any form of psychological treatment must be applied early on, before the victim's immune system becomes crippled beyond salvation, if it is going to improve the physical condition of a person with HIV. In practice, this means not long after the initial infection and well before the first symptoms of AIDS appear. For someone who already has full-blown AIDS, it is probably too late for treatments such as stress management, relaxation, group therapy or physical exercise to have any useful effects on their health (though these techniques can still do much to ease the victim's psychological distress).

The theoretical potential exists for psychoneuroimmunology to help, but this potential is, as yet, unrealized. Psychological interventions can improve the mental wellbeing and quality of life of those with HIV or AIDS. Whether such interventions help them to live longer remains to be seen.

Imagery, miracle cures and other exotica

> Sorcerers are too common; cunning men, wizards, and white witches, as they call them, in every village, which, if they be sought unto, will help almost all infirmities of body and mind.
>
> Robert Burton, *Anatomy of Melancholy* (1621)

> . . . these smouldering village beliefs made him furious if mentioned, partly because he half entertained them himself.
>
> Thomas Hardy, 'The Withered Arm' (1888)

Now we journey into wilder spheres, away from the reassuringly scientific domain of psychoneuroimmunology and into the realms of those latter-day wizards, the New Age gurus. The comforting notion that life-threatening diseases can be cured by purely non-physical means is as old as religious belief itself. Long before the world's monotheistic religions developed, humans believed in the supernatural power of disembodied gods, ghosts, demons and spirits to cure (or cause) physical ailments. The once-common faith in the curative powers of touching a freshly hanged man, for example, forms the core of 'The Withered Arm', one of Thomas Hardy's tales of life in nineteenth-century rural England.

Gertrude Lodge is afflicted with a withered arm. In desperation she seeks the advice of the local quack, Conjuror Trendle, who recommends that she 'touch with the limb the neck of a man who's been hanged . . . Before he's cold – just after he's cut down . . .' as this will 'turn the blood and change the constitution'. Alas for poor Gertrude, there are unforeseen complications and she ends up dead.

Tens of thousands of people each year make the pilgrimage to holy shrines like Lourdes, expecting to be cured of their physical ailments by divine intervention. Many believe that they have indeed been cured, although the Catholic Church accepts only a tiny number of such claims as true miracle cures. A scientific review of Lourdes cures found that of the six thousand or so well-documented cases, only sixty-four were sufficiently convincing to fulfil the Vatican's strict criteria for a miracle cure, and a number of these involved diseases that are known to show occasional spontaneous remission.[4]

Cancer, that most terrifying of diseases, has provided fertile territory for a swelling band of New Age gurus, offering appealingly simplistic explanations and latter-day miracle cures. One of their clarion calls is that cancer starts in the mind and can therefore be cured by the mind.

In the previous chapter we saw striking evidence that supportive social relationships can help to prolong the survival of cancer patients, as well as greatly improving their quality of life. But the well-founded idea that a patient's mental state can influence the progression of their disease has been taken several stages further by the New Age gurus. Cancer, they assert, stems from thinking the wrong thoughts or not loving yourself enough. Hence, salvation lies in using the power of the mind to destroy the cancer. It is all a question of having the right attitude and enough love; think positive and you will be well again.

Proponents of these seductively comforting ideas have made many thousands of converts throughout the world. Unfortunately, on the rare occasions when scientists have been granted the opportunity to assess their claims objectively, they have generally been found to have little validity.

Another vogue idea is that cancer victims can fight their disease by imagining their white blood cells attacking and destroying the cancer cells – a technique known as visualization or guided mental imagery. Some people report that this technique has helped to bolster their sense of psychological wellbeing, especially when combined with relaxation techniques. There is evidence that a blend of mental imagery, relaxation and biofeedback can be used to modify the numbers of circulating white blood cells, natural killer cell activity and other aspects of immune function. When Danish researchers at the University of Aarhus asked healthy volunteers to use guided imagery and relaxation techniques in a conscious effort to boost the effectiveness of their immune systems, it brought about a significant increase in the volunteers' natural killer cell activity over a ten-day period. Notwithstanding intriguing results like this, the belief that guided imagery can by itself cure cancer rests more on hope than on science.

A traditional panacea that does seem to work is simply getting things off your chest – or self-disclosure, to use the weightier psychological parlance. Since the dawn of civilization, humans have found that unburdening their woes, anxieties or traumas to a sympathetic listener usually makes them feel better. In addition, self-disclosure often helps us to formulate solutions to our problems. Talking to a skilled listener is, of course, the basis of most types of counselling and a central element in psychotherapy. There is little doubt that the appropriate use of self-disclosure can improve our psychological and emotional state in troubled times. But can it also improve our immune function or physical health? There are hints that it might.

Janice Kiecolt-Glaser's group asked a sample of healthy student volunteers to write about emotionally charged, upsetting events in their lives. The students were, in effect, asked to unburden themselves on paper. Those who took the experiment seriously and confronted their past traumas were found to have more responsive lymphocytes than a comparison group who had been asked to write about impersonal matters. Furthermore, the self-disclosing students made fewer visits to the health centre over the following weeks, implying that they were either healthier or less neurotic (or both).

Similar findings came from a research project conducted at the University of Miami. Students were asked to recall and discuss stressful or traumatic events in their lives which they had previously kept to themselves. Individuals who revealed the most about themselves, or recalled particularly troublesome events, showed better immune function, as evidenced by immune control over herpes viruses.

All in all, it would seem that the emotional relief of unburdening oneself can have beneficial effects on the body as well as the mind.

Kill or cure?

> As if you would call a physician, that is thought good for the cure of the disease you complain of but is unacquainted with your body, and therefore may put you in the way for a present cure but overthroweth your health in some other kind; and so cure the disease and kill the patient.
>
> Francis Bacon, *Essays* (1625)

Many of the wilder claims about the power of the mind to cure serious diseases are built on foundations of sand. But some of the New Age gurus' therapies are worse than merely ineffectual. They, and the philosophies that underpin them, do more than just waste people's time and money: in some circumstances they are positively harmful.

The extreme view that all disease originates in the mind implies that sick people ultimately have only themselves to blame for their disease or their failure to recover. If you can cure yourself by thinking the right thoughts, or harbouring the right emotions, then you must have fallen ill in the first place because you were thinking the wrong thoughts or harbouring the wrong emotions. Even more corrosive is the insinuation that if your child gets ill then you, its parents, must be to blame for failing to give enough love.

The belief that physical illness can have purely mental or emotional origins imposes an additional burden of guilt on people who may already be seriously ill. This guilt can, on occasion, prove a greater problem than the disease – and far harder to treat. Patients with life-threatening conditions such as cancer are exceptionally vulnerable to feelings of guilt and depression. To imply that their disease is the product of their own incorrect thinking is far from helpful. It also happens to be untrue – or, at best, a grossly inadequate version of the truth. Yet there are practitioners of the New Age gospels who explicitly preach this insidious message when they proclaim, for example, that people get cancer because they do not love themselves enough.

One reason why this flawed philosophy has a superficial appeal is its perverse implication of justice: we get the diseases we deserve; you are ill so you must have been doing something wrong. There are strong echoes here of old religious doctrines that looked upon disease as a divine punishment for sin or moral lapse. The assumption that someone must be to blame for the disease (even if it is the victim) resonates with the consumerist culture of modern industrialized societies; we all have choices and it is our fault if we make the wrong ones.

Robert M. Sapolsky, a leading researcher on the biology of stress at Stanford University, admirably encapsulated the views of many scientists (and my own) when he wrote:

> It is bad enough to have cancer without being led by some perversion of psychoneuroimmunology into thinking that it is your fault that you have it and that it is within your power to cure it.

An undiscriminating faith in the magical curative power of pure thought has other, not inconsiderable, hazards as well. Most dangerously of all, it can incite patients to reject the conventional medical therapies on which their lives may depend. Conventional medicine might have its blind spots and weaknesses, but it still offers the best hope of survival for those with life-threatening diseases such as cancer.

But even conventional medical treatments can have hidden dangers if, as often happens, their psychological and emotional consequences are ignored. Telling a person that they have a high risk of developing cancer or heart disease will inevitably subject them to psychological stress. Over the years, psychologists have accumulated a great deal of information about how people react when told that they have a serious disease or a high risk of developing one. Many individuals cope robustly with the bad news and soon master their anxieties. But others harbour continually intrusive worries about their condition. They are, in effect, experiencing chronic stress. Even something as innocuous as attending a routine health screening may prove stressful for some individuals; psychological studies have found that a routine screening for coronary heart disease can provoke substantial anxieties that last for months.

You will be unsurprised to learn that if this sort of stress is severe enough it can have adverse effects on physical health, as well as causing emotional distress. Discovering that you are seriously ill, or that you have a high risk of developing a serious illness, is in itself stressful and could potentially make your condition worse. Paradoxically, the mere act of informing a person that they are at high risk can, in some instances, increase that risk.

Experiments conducted at the University of Oslo demonstrated that telling people they had high blood pressure increased their resting blood pressure still further and made them more reactive to stressors. Young men with hypertension who were made aware of their condition were later found to have higher resting blood pressures than hypertensives who remained unaware of their condition. They also showed larger increases in heart rate and adrenaline levels when subjected to a mild physical stressor.[5]

Now, I am certainly not suggesting that doctors should avoid telling us the unwelcome truth about our health in case it upsets us. We all have a right to know about our health, and we positively need to know if we are going to take appropriate remedial action. A person who remains blissfully ignorant of their hypertension will have less cause to give up smoking or avoid overworking. There is no easy solution to this dilemma. However, there clearly is a real need for doctors to be sensitive to the psychological, emotional and physical consequences of giving people disturbing information about their state of health.

Contrary to the claims of certain self-help manifestos, there are no quick fixes or easy solutions to problems like cancer or heart disease. It is probably safe to conclude that having the right attitudes to life, a satisfying job and a network of supportive personal relationships is of greater benefit than any of the esoteric therapies or self-help remedies on offer.

Psychoneuroimmunology does not offer us magical cures for sickness and unhappiness (not yet, anyway). So far, it has generated surprisingly few new solutions to medical problems. However, this is a new area of science and it takes time for theoretical and experimental knowledge to be translated into practical remedies. Like a young child, psychoneuroimmunology has enormous potential which is as yet largely unfulfilled. But psychoneuroimmunology does at least offer us solid knowledge and understanding, and without these we would inevitably remain the prisoners of our ignorance.

11

Exorcising the Ghost in the Machine

> All the mystery in life turns out to be this same mystery, the join between things which are distinct and yet continuous, body and mind . . .
>
> Tom Stoppard, *Hapgood* (1988)

> [There is no] Ghost in the Machine
>
> Gilbert Ryle, *The Concept of Mind* (1949)

If the connections between our mental state and our physical health are so real and so fundamental then why all the fuss? How come psychoneuroimmunology is such a shiny new subject instead of a venerable academic tradition? The odd fact is that until quite recently the connections between mind, body and health have been remarkably neglected by science and medicine. Why?

Historically, the study of physical disease processes (pathology, immunology and allied disciplines) developed separately from the study of the mind and behaviour, and never the twain did meet. To most immunologists the mind was an irrelevance, while few psychologists or psychiatrists gave much thought to lymphocytes or bacterial infections. Medicine was concerned with mechanical defects in the physical body. Thoughts, emotions and other ephemera were, both literally and metaphorically, immaterial.

In the past, immunology dealt with the immune system as though it were an autonomous, self-regulating entity which operates independently of the mind. When challenged, immunologists pointed to the fact that many basic immunological reactions can be made to occur in the test tube, far removed from nerve cells. When confronted with claims about the apparent influence of psychological and emotional factors on immunity, the sceptics argued that even if the brain could affect the immune system, its impact would be too feeble to make any tangible difference to anyone's health. Some continue to hold this view. But most simply ignored the mind altogether.

On the opposite side of the intellectual chasm, many psychologists, psychiatrists and practitioners of psychosomatic medicine adopted an equally lop-sided view. They readily accepted the separation between organic diseases and psychosomatic diseases, with its dubious assumption that bodily disorders are either physical (in other words, real) or all in the mind. Their attention was focused on the mind (or, occasionally, the brain) and they had little interest in the base mechanics of the body.

The underlying split between mind and body, ethereal and physical, also applied to the study of mental illness. Historically, a distinction was drawn between diseases of the mind, which were the province of psychiatry and psychology, and diseases of the physical brain, the province of neurology and neurobiology.

The state of science, then, was one of almost complete separation between mind and body. It still is, to a considerable extent. Of course, there have always been honourable exceptions to this rule – immunologists who thought about psychological factors and behaviour, and psychologists who wondered what the mind could do to immunity and health – but they represented a small minority.

Fortunately, as Aldous Huxley once observed, 'facts do not cease to exist because they are ignored'. Now we know better. With the belated advent of psychoneuroimmunology, the realization is at last beginning to dawn that the mind and the immune system do not operate independently of one another. But why have science and medicine been so polarized in the way they have habitually regarded the mind and body as two completely different sorts of thing? The roots to this problem run deep and its tendrils extend throughout the heritage of Western thought.

That old mind–body problem

For of the soul the body form doth take;
For soul is form, and doth the body make.

Edmund Spenser, *An Hymn in Honour of Beauty* (1596)

It is certain that this 'me' (that is to say, my soul by which I am what I am), is entirely and absolutely distinct from my body and can exist without it.

René Descartes, *Meditations VI* (1641)

The modern scientific debate about the role of psychological and emotional factors in disease is in many respects a manifestation of an older and wider philosophical debate: the mind–body problem.

Among the most enduring of all philosophical issues, the mind–body problem continues to cast a pervasive influence over science, medicine and the arts. At issue is the nature of the relationship between mind (thoughts, feelings, emotions, perception, consciousness) and matter (the body and brain). Are the two wholly distinct sorts of entity or, conversely, different aspects of a single entity?

Dealing as it does with the very nature of consciousness, thought and human existence, the mind–body problem is far more than just an abstract philosophical debate. It goes to the very heart of how we conceive of ourselves and the world around us. Scholars have been pondering on the mind–body problem for as long as there have been scholars to ponder. At the risk of being grossly simplistic, it is helpful to divide opinion into two distinct camps – dualism and monism.

The school of thought that regards the mind and body as two intrinsically separate concepts is called dualism.[1] Dualists reject the notion that the workings of the mind can be explained solely in terms of electrical and chemical processes within the physical brain. Hard-line dualists would maintain that, in principle, minds can exist without bodies and bodies can exist without minds. (Perhaps you know someone who exemplifies the latter category.)

The contrasting belief in the basic unity of mind and body is called monism.[2] According to this viewpoint, the mind and mental processes can be explained – in principle if not yet in practice – in terms of physical processes occurring within the physical brain and body. A true monist would recoil at the idea of a disembodied, non-physical mind, spirit or *psyche*. The British philosopher Gilbert Ryle once described the dualist concept of the mind as a non-physical entity as the 'dogma of the Ghost in the Machine'.

For centuries dualism has prevailed in Western thought. The English language embodies this dualist split: we instinctively contrast mind and body, body and soul, mental and physical, *psyche* and *soma*, mind and matter, flesh and spirit, as naturally as we compare night with day.

This philosophical heritage of mind–body dualism still permeates science, medicine and the arts, even as we enter a new millennium. Dualism lives on, for example, among those Artificial Intelligence experts who believe that the human mind is like a vast computer program, while the brain is the biological hardware on which the mind software happens to run. The dualist assumption buried at the heart of this computer analogy is that the mind and body can be understood without reference to one another, in the same way that computer software can be understood without reference to the particular machine on which it operates.

It is natural that a debate which penetrates to the core of human nature should exercise the minds of writers and poets as well as philosophers and scientists. William Blake savagely rejected the establishment view that the mind, or soul, is distinct from the physical body. In *The Marriage of Heaven and Hell*, written in 1790, Blake attacked the accepted values of his age by asserting the unity of body and soul:

> All Bibles or sacred codes have been the causes of the following Errors.
>
> 1. That Man has two real existing principles: Viz: a Body & a Soul.
> 2. That Energy, call'd Evil, is alone from the Body; & that Reason, call'd Good, is alone from the Soul.
> 3. That God will torment Man in Eternity for following his Energies.

> But the following Contraries to these are True:
> 1. Man has no Body distinct from his Soul; for that call'd Body is a portion of Soul discern'd by the five Senses, the chief inlets of Soul in this age . . .

A century later, Charles Darwin expressed a scientific brand of monism when he wrote:

> Why is thought being a secretion of the brain more wonderful than gravity a property of matter?

Darwin's implicit rejection of mind–body dualism is in tune with what twentieth-century science has discovered about the brain. Science has become increasingly able to provide satisfying explanations of thought, emotions and consciousness in terms of physical processes within the brain. As scientific understanding of the physical brain has expanded, so the need to assume the existence of a non-physical mind, hovering inside the body like a disembodied spirit, has diminished.

Many scientists and philosophers are now willing to regard the mind as a product or manifestation of the physical brain. Few are willing to accept that minds can exist separately from physical bodies, even in principle. Indeed, a growing number of biologists are taking this rejection of mind–body dualism a stage further. Research in the neurosciences has stressed the utter dependence of the mind upon the body. The brain and the rest of the body form an integrated organism. The body provides far more than just a life-support system for the brain, and a disembodied brain would not constitute a mind. Yet scientists are in a small minority.

But what has all this got to do with psychoneuroimmunology and health? Mind–body dualism fosters the belief that psychological and emotional factors have little or nothing to do with immunity or disease. The mind and body might occasionally impinge upon each other in certain ways, but they are essentially separate and work according to different principles. Dualism provides philosophical succour to the sceptics and Roundheads who prefer to exclude psychology and mental phenomena from any analysis of physical diseases. Dualism and psychoneuroimmunology might not be entirely incompatible, but neither are they natural bedfellows.

THE GRIP OF DUALISM

Why has this dichotomy between mind and body permeated Western thought for so long? I cannot pretend to offer a firm answer, but I shall hazard a few tentative suggestions why the influence of mind–body dualism has been so enduring.

First and foremost, dualism has a strongly intuitive appeal. Minds and bodies *seem* like very different sorts of beast, especially to someone who knows nothing about the workings of the brain. The attributes we associate with the mind – our consciousness, thoughts, memories, emotions and human individuality – appear to be wholly separate types of phenomena from the mechanical functionings of the body. The body is like a machine; the mind is not.

Our mental life is a uniquely private and subjective experience: you can objectively measure the electrical activity in my brain or the hormone concentrations in my blood, but you cannot have any direct knowledge of what is going on in my mind (and probably just as well). Minds – unlike bodies or machines – appear to be imbued with a sense of purpose and intentionality. When we decide to do something it is our mind that makes the decision; our body merely does what it is told. Whatever the true natures of the mind and body, we need different terminologies to discuss them. The vocabulary of chemistry or neurobiology is of no practical use when we want to describe our thoughts, feelings or intentions.

Humans have long presumed that our conscious minds are what distinguish us from mere machines, and also what distinguishes one individual from another. Your mind is what makes you both distinctly human and distinctly you. Your brain, on the other hand, is merely a porridgy lump of wet organic computer. It is difficult to imagine how this unprepossessing lump of meat, weighing about the same as a bottle of champagne, could possibly be the seat of consciousness, emotions, individuality and creativity.

And long before people ever thought about machines, the conscious, reasoning mind was what supposedly distinguished humans from other species. Indeed, some scholars have suggested that a deep-seated need to distinguish humans from the 'lower' animals has strengthened the tenacious grip of mind–body dualism. To most people, even today, the mind is far more than just a brain. Monism might be more compatible with modern scientific knowledge than dualism, but it is not compatible with our common perceptions.

An ancient doctrine that highlighted the dualist split between body and mind was the notion that illness is sent by the gods to punish humankind for moral lapses. In *The Odyssey* and *The Iliad*, for example, Homer uses the idea of illness as divine punishment. In the ancient cultures of Assyria and Babylonia the same word was used to denote both disease and sin. Symptoms of disease were regarded as the physical manifestations of an underlying moral disorder. Diagnosis involved trying to discover what sin or moral offence the sufferer had committed. Treatment then consisted of performing expiatory rites or sacrifices to placate the angry gods and atone for the sin. From the patients' point of view, faced with healers who lacked any understanding of how bodies work, this was probably the least harmful option.

In case you think the doctrine of disease as a punishment for sin died out centuries ago, rest assured that it is alive and kicking. It revealed itself in the 1980s when modern proponents preached the message that AIDS was a punishment for the unnatural and sinful behaviour of its victims. As one senior British policeman said, in a 1986 speech referring to those most at risk from AIDS: 'Everywhere I go I see increasing evidence of people swirling around in a human cesspit of their own making.'

Monism, by contrast, is uncomfortably at odds with religious philosophies which emphasize the distinction between body and soul, flesh and spirit – a dichotomy which has been a feature of most religions since the dawn of civilization.

As far as archaeologists and historians can deduce, humans have harboured beliefs about non-physical spirits, souls, gods, ghosts and demons ever since we evolved the capacity to think. Archaeological evidence from cave paintings and ancient funeral sites suggests that people believed in the existence of disembodied spirits twenty thousand years ago. One ancient form of medical treatment involved boring holes in the skulls of mentally disturbed individuals to release the evil spirits, a rather alarming practice known as trephination.

Dualism is inherent in the idea that we continue to exist in some form after our physical bodies die. To believe in the existence of disembodied spirits, eternal souls, life after death or ghosts is to take a dualist stance on the mind–body problem. To reject dualism is to reject the comforting thought that human existence transcends the physical and can endure without it. If you maintain that your mind is a product of your brain and inseparable from it, you must accept that when your body dies you die along with it. The death and decomposition of physical bodies is an observable fact, but what happens to the mind or soul after death is still a subject of debate and faith.

No wonder dualism has been the predominant perspective in Western thought and remains so to this day.

SOME ANCIENT HISTORY

Dualism has not always maintained an unwavering grip on Western thinking, however. An element of monism did creep in for a while. To ancient Greek philosophers the distinction between mind and body was not as absolute as it subsequently became. Ambivalent perspectives can be found in the works of Plato and Aristotle, amongst others.

Early exponents of views that could properly be called monist included the Greek scholars Alcmaeon, Epicurus and Hippocrates. They questioned established beliefs about disembodied spirits and instead offered explanations for the natural world, including the human body, in terms of observable, physical phenomena. Hippocrates, who was born in about 460 BC and whose name lives on in the eponymous medical oath, was the most prominent physician of his time and earned the epithet 'the father of medicine'. He sought to explain the workings and defects of the human body in essentially mechanistic terms, rather than resorting to the usual panoply of gods and evil spirits.

Central to Hippocratic medicine was the tenet that health depended upon a harmonious balance of the humours (blood, phlegm, yellow bile and black bile) and that diseases were caused by an imbalance in these vital fluids. Hippocrates rejected the traditional creed that disease was caused by spiritual or supernatural agents which disturbed the body's harmony. As far as he was concerned, diseases were caused by natural forces which could be studied and explained. The same was true for the body's reaction to these disease-causing forces. Hippocrates argued that disturbances in the body's internal harmony provoked adaptive responses which tended to counteract the disturbance and return the body to its state of healthy harmony. Nature, rather than the gods, was both the source and the healer of diseases.

Though technically wide of the mark by modern standards, Hippocrates' approach to medicine was revolutionary. By focusing on the fleshly realities of diseases, instead of their supposedly divine or spiritual origins, Hippocrates and his followers were able to make substantial progress in understanding and treating real illnesses. In contrast, the traditional approach of attributing illness to angry gods or evil spirits was hardly conducive to scientific enquiry, let alone systematic remedies. The philosopher Epicurus, who was born in 341 BC, developed Hippocrates' ideas about disease even further. Foreshadowing in some respects the basic concept of psychoneuroimmunology by over two thousand years, Epicurus argued that the mind is one of the adaptive forces which influences the body's health and its response to physical disease.[3]

Greek medicine had reached its zenith by the second century AD. The foremost medical authority in the Western world at this time was the Greek physician Galen, who was born in about 130 AD. After starting his medical career as a physician to the gladiators of Pergamum in Asia Minor, Galen moved to Rome and became court physician to the emperor Marcus Aurelius. He made a number of significant medical advances, including establishing the use of the pulse as an aid to diagnosis, although he also got a lot of things embarrassingly wrong.

The teachings of Galen reflected the earlier ideas of Alcmaeon, Hippocrates and Epicurus. They emphasized a rational, materialistic approach to the understanding and treatment of disease. To Galen and his disciples, diseases were abnormal physical conditions of a physical body and not aberrations of the soul or punishments meted out by an angry deity. Their materialistic approach to medicine acknowledged the existence of the *psyche* (the spirit, soul or mind, depending on how you translate it) but in practice paid it scant attention. The focus of their medicine was almost exclusively on the mechanics of the physical body, or *soma*. Wind the clock forward two thousand years and not much has changed in this respect.

The surprisingly up-to-date theories of mind, body and disease that flourished in ancient Greece did not, however, endure. They were soon swamped by ecclesiastical doctrine and a return to the orthodoxy of dualism. Not long after Galen's time, Greco-Roman civilization began to collapse and Europe sank into the intellectual abyss of the Dark Ages. During the centuries that followed, Western science and philosophy were the prisoners of dogma, and medicine became fossilized.[4]

The early Christian scholars steered a middle course between the medical theories of the ancient Assyrians and Babylonians, who regarded disease as divine punishment for sin, and the materialistic medicine of Greek scholars like Hippocrates and Galen, who saw diseases as physical disorders of the body. To the early Christians, sin was not in itself a direct cause of disease, although it could indirectly give rise to disease.

By the Middle Ages, the prevailing Western attitude towards disease had reverted to a state of undiluted dualism and the separation of mind from body was back to stay. Medieval scholars adapted the philosophical texts of Aristotle and his peers to fit into the Christian framework, reconciling faith with reason.[5] The human mind or soul was explicitly viewed as an immortal, non-physical and God-given entity, completely different in kind from the material body. The medical writings of Galen were incorporated into this Christianized version of Greek philosophy and became the absolute authority on all medical matters for centuries to come. The hallmarks of Western scholarship during this period were an unquestioning obedience towards established authority, the supremacy of ecclesiastical doctrine and an obsession with arcane irrelevancies. These were the men who debated how many angels would fit on the head of a pin.

Belief in the unalloyed validity of Christianized Greek medicine did not crumble until the fifteenth and sixteenth centuries, when Renaissance scholars began to reject the hallowed doctrines and a rational, scientific approach to medicine was gradually reborn. By the seventeenth century the stage was set for a truly scientific approach to medicine. Galen's teachings about the blood and heart were shown to be spectacularly wrong when the English physician William Harvey (1578–1657) discovered how blood circulates within the body.

Paradoxically, though, it was in the seventeenth century that mind–body dualism tightened its grip on Western thought and ensured its dominance for another three hundred years. This came about largely through the works of one man: René Descartes.

René Descartes and the separation of mind from body

René Descartes, the seventeenth-century French philosopher and mathematician, probably did more than any other individual to formalize and entrench the dualist split between mind and body in contemporary Western thought. Indeed, the intimate association between Descartes and dualism is embodied in the term 'Cartesian dualism'.

From Descartes' time onwards, a belief in the mind (or soul) as an intangible, non-physical entity, distinct from the physical body, became the pre-eminent view in Western thought. (Though then, as now, an 'unofficial' medicine continued to thrive, in which the patient's mind was as much the practical focus of treatment as their physical body.)

In Descartes' philosophy, the human mind or soul (the two were synonymous) was not a physical object that could be located in space. Rather, the mind was a non-physical entity, implanted in the human body by God. The mind, or *res cogitans* ('thinking thing'), was responsible for conscious thought, and conscious thought was the very essence of human existence. *Cogito, ergo sum* – I am thinking, therefore I exist. Conscious thought was what made humans different from other species. (Descartes was adamant that non-human species did not have minds: animals were merely clever machines. This credo was later used by vivisectionists in the seventeenth and eighteenth centuries to justify various barbarities.)

In the world according to Descartes, the mind is in no sense a product of the brain or any other physical structure; it 'needs no place and depends on no material thing'. Moreover, it can exist independently of the body and leaves the body when we die. Thus far Descartes' philosophy was consistent with religious doctrine.

The fact that the mind and body patently do affect each other all the time was a source of considerable difficulty for Descartes' theories. How could an ethereal mind, with no physical location or substance, impinge upon a fundamentally different kind of entity like the physical body? Descartes never quite got around this one. He did, however, nominate a precise physical location where the interaction between mind and body was supposed to occur – namely, the pineal gland or 'liaison brain'. (One of the many popular fallacies about Descartes is that he identified the pineal gland as the physical location of the soul. He was subtler than that.)

By presenting the world with a philosophical system which reconciled religious belief with scientific reason, Descartes formalized the division between mind and body and lent such force to the dualist argument that its grip over Western thought lasted for the next three hundred years. It was this dualist approach that led ultimately to the woeful neglect of the mind by modern medicine and the deep-seated scepticism about the role of psychological and emotional factors in disease. The dualist split created the vacuum in medicine which holistic 'alternative' forms of medicine have expanded to fill. It also served to bolster the psychosomatic fallacy, whereby diseases are regarded as either 'physical' (i.e., real and not the victim's fault) or 'psychological' (i.e., not real and implicitly the product of personal weakness).

Descartes' dualist legacy has encouraged scientists and philosophers to neglect the emotions and place undue emphasis on conscious, rational thought. Modern neuroscientists are only now beginning to realize that the emotions are key ingredients of rationality and vital for normal human functioning. The Cartesian notion of a disembodied, rational mind, unencumbered by mere emotion, simply does not correspond with the true nature of the human organism. We are no more capable of coping with life solely by rational calculations than we could through emotions and feelings alone. Both are essential if we are to function in the real world. Prominent neurobiologists such as Robert Damasio and Gerald Edelman have advanced detailed and compelling arguments why the mind is dependent on the whole body, not the brain alone. They have successfully put emotions back on to centre stage.

Notwithstanding this catalogue of criticisms, it would be quite wrong to portray Descartes as a philosophical villain who delayed the advent of psychoneuroimmunology by centuries. If we examine the balance sheet we find that Descartes' enormous contributions to the progress of science and medicine far outweigh his dualist foibles. To understand the full extent of Descartes' achievements we must put things in historical perspective.

Descartes was a sceptic who searched for rational principles to explain the mysteries of the natural world. In doing so, he came perilously close to rejecting certain sacred tenets of ecclesiastical dogma. His method was to analyse the human body in much the same way as an engineer would analyse the workings of a mechanical device. He applied the full armoury of mathematics and physics to the functioning of living organisms. By making this scientific approach philosophically acceptable, Descartes was instrumental in overturning the dogmas that had stultified Western thought for over a millennium. Thus he banished the sterile notions of life forces, spirits and souls and turned the spotlight on the physical realities of living bodies.

In addition, he stripped away many of the former territories of the immortal soul and reduced its boundaries to the bare minimum: the functions of consciousness and rational thought. The rest of human nature, including perception and the emotions, he placed firmly within the realm of the physical body. To Plato, Aristotle and most Western thinkers before Descartes, the very essence of life itself had resided within the soul. To Descartes, however, life was a property of the physical body and only human consciousness resided in the mind. Emotions and feelings could be explained in terms of organic secretions within the body. This was revolutionary stuff.

In the seventeenth century this scientific approach to human nature was regarded by the authorities as a threat to established religious belief and the power of the church. Most of us are fortunate to live now in a world where we are reasonably free to express our thoughts, explore new ideas and question dogmas. We should not forget that the world was a much less tolerant place in Descartes' time.

In 1596, the year of Descartes' birth, William Shakespeare was just getting into his stride as a playwright, Galileo invented the thermometer and the first ever water closet was installed. Barely fifty years had elapsed since the Polish astronomer Copernicus had first proposed the revolutionary theory that the earth is not the fixed centre of the universe.

In Europe at this time anyone who presented a serious challenge to religious orthodoxy faced torture or execution at the hands of the Holy Office, better known as the Inquisition. When Descartes was four years old the Italian philosopher and scientist Giordano Bruno was burned at the stake in Rome for daring to espouse the heretical theory that the earth rotates on its axis and orbits around the sun. In 1633, when Descartes was thirty-seven years old, he heard that the great Italian scientist Galileo had been denounced by the Inquisition and forced, under threat of torture, to recant his scientific theories. Galileo's great crime, like Giordano Bruno's, had been to agree with Copernicus that the earth is not the centre of the universe. Fearing for his own safety, Descartes deliberately withdrew from publication one of his own pro-Copernican books.

Descartes' writings pushed at the boundaries of acceptability and were as daring as the times would allow. Like Copernicus, Bruno and Galileo before him, Descartes was convinced that the accepted dogma of the earth as the centre of the universe was plain wrong. The astronomical evidence showed that the earth was spinning daily on its axis and orbiting annually around the sun. But this was not what the church authorities wanted to hear, as Bruno and Galileo found to their cost. Descartes' belief that the human body could be regarded as a complex machine was equally controversial.

Painfully aware of Galileo's fate at the hands of the Inquisition, Descartes wanted only to live in peace, without attracting any hostile attention to himself or his works. He remained a Catholic all his life and was careful not to push his luck. His philosophical system scrupulously preserved the essential notion of the immortal, God-given soul. Even so, he won no plaudits from the church and in 1663 Rome declared his works to be damnable.

It is not surprising that Descartes' more theologically challenging writings were subtly coded and won him the epithet of 'the masked philosopher', nor that his tombstone carries the telling inscription *Bene qui latuit, bene vixit* ('He who hid well, lived well'). Now, I am certainly not implying Descartes was a closet agnostic. On the contrary, the existence of a benevolent God was absolutely central to his philosophical system. However, some of his writings did sail perilously close to the wind as regards the doctrines of the all-powerful church, and Descartes would have felt that chill wind on his neck more keenly than most.

Before Descartes' time, Western medicine had been kept in suspended animation. Orthodox beliefs had maintained that the human body was a sacred object and therefore out of bounds for scientific enquiry. For all its negative ramifications, the way in which Descartes separated the mind from the body, and made this separation theologically acceptable, was a tremendous blessing for medicine. After a thousand years of intellectual stagnation, Cartesian dualism helped to free scientists to recommence their investigations of the human body, heralding the dawn of scientific medicine. Dissecting a human corpse was no longer seen as a violation of the sacred spirit. Experiments in pursuit of medical knowledge were legitimized. Thus Descartes liberated science from the shackles of dogma. The price – including the neglect of the mind in modern medicine – was probably worth paying.

DESCARTES' DEMISE

Although he is best known for his mathematical and philosophical works, Descartes was keenly interested in medicine throughout his life and his writings reflect this. Indeed, in *Meditations* he relates how, as a sickly child, he personally experienced the power of the mind when he conquered a life-threatening illness:

> I believe that the inclination that I have always had of con-

> sidering things which come up from a point of view which could enable me to render them most agreeable, and to make sure that my principal contentment depended on me alone, is the reason that this indisposition which came to me as if it were natural to me, has little by little entirely passed away.

There is more than a hint of the young Pollyanna in this statement. Descartes had clearly discovered the benefits of adopting a positive attitude.

Ironically, psychological stress may have contributed to the eventual demise of the great dualist himself. In September 1649 Descartes left Holland, where he had been mired in academic controversy, to seek a quieter life as the personal tutor to Queen Kristina of Sweden. Unfortunately, he stepped out of the intellectual frying pan and into the psychological fire.

Life in Queen Kristina's court might have been far removed from the intellectual battles that had so wearied Descartes, but in other respects it was much harsher. After a depressing winter engaged in pointless royal fripperies, Descartes finally got down to work as a tutor in January 1650. It so happened that Queen Kristina believed her intellectual powers were at their peak at five o'clock in the morning.[6] Thus it was at that unearthly hour that the unfortunate Descartes was obliged to instruct her in philosophy. Like many eminent scholars, he had been used to a more congenial lifestyle. He was not an early riser and had been accustomed to lying in bed until late in the morning, pondering. Even as a schoolboy he had been granted special dispensation to get up late.

The discomfort and stress of this dramatic change in his lifestyle, combined with the freezing Stockholm mornings, proved too taxing for Descartes' mind and body. Within a month of starting this harsh new regime, shortly before his fifty-fourth birthday, he took ill and died – probably of pneumonia.[7]

Descartes' last recorded words had a distinctly dualist flavour: '*Ça mon âme, il faut partir*' ('So my soul, a time for parting').

12

A Fresh Pair of Lenses

For a biologist the alternative to thinking in evolutionary terms is not to think at all.

P. & J. Medawar, *The Life Science* (1977)

Is there an alternative lens through which we can examine the myriad links between brain, behaviour, immunity and disease? Modern medical science has been enormously effective in analysing the mechanical workings of the human body, but it has paid much less attention to questions like why our bodies are designed in this way and why certain individuals are prone to ill health. Biologists, on the other hand, are used to asking such questions.

Conventional medicine is primarily concerned with analysing how our bodies operate, so that their defects can be understood and rectified. Medical scientists seek to understand the genetic, biochemical and physiological mechanisms of our bodies in much the same way that an engineer who wants to repair a broken machine is concerned with how its cogs and gear-wheels work. This is a perfectly sensible strategy, and it has proved highly effective in generating solutions to medical problems. But it does not provide the complete picture. Examining old medical problems from new perspectives can often prove fruitful.

In this final chapter I want to describe two new ways of looking at the connections between mind, body and disease. They are the biological perspectives of development and evolution.

Scientists can ask other sorts of question besides 'How does it work?' An equally valid question is 'How did it develop?' – that is, how did the biological structure or behaviour pattern arise during the lifetime of the individual, through the interplay of genetic and environmental influences. When biologists investigate the workings of adult organisms they are mindful that each of those organisms (including you and I) started life as a microscopic and apparently formless single cell. We did not spring into life as fully fledged adults. Our imaginary engineer would be asking how the machine was assembled during the manufacturing process.

Evolutionary origins and biological functions provide another field of enquiry. Here, the focus is on why things are the way they are, rather than how they work or how they were assembled during development.[1] Evolutionary questions deal with how biological structures and behaviour patterns have been shaped by the processes of natural selection over the evolutionary history of the species, and how these structures and behaviour patterns help the organism to survive and reproduce in its current environment. Our imaginary engineer would be thinking about what the machine is for and why it has been designed in that particular way. Our imaginary doctor would be asking why, rather than how, we become ill.

Development

> The Child is father of the Man
>
> William Wordsworth, *My Heart Leaps Up* (1807)

The study of development tackles such fundamental issues as how genes and environment interact to produce a unique individual, and how an organism's characteristics change (or remain resistant to change) during the course of its lifetime, from conception to death.[2] Thus it encompasses the staggeringly clever processes that transformed you from a single cell, visible only under a microscope, into the unique human being you have presumably become.

Development is a truly extraordinary process. Each of us starts life as a single fertilized cell containing 46 double strands of DNA, along which are arranged approximately 100,000 genes. Years later, by means that are as yet only dimly understood, we each end up as a distinctive individual with a body consisting of over ten million

million (10^{13}) highly specialized cells. Each of those ten million million cells is in the right place and doing the right things, and as a result each one of us is unique. You are different from me in many ways (for which you should probably be thankful).

Unravelling the interacting contributions of genes and environment during development (the 'nature–nurture problem') is not a straightforward matter. The consequences of inheriting particular genes from your parents will depend on your environment, while the effects of environmental factors will in turn depend on your genetic make-up. Take the question of disease-prone behaviour patterns, for example. Why do some people take up smoking or develop highly reactive Type A personalities, making themselves vulnerable to coronary heart disease? The roots of these individual differences in behaviour often lie very early in life, and reflect both our genetic inheritance and personal experience.

Social pressures and learning clearly play important roles in health-related behaviour patterns such as smoking and drinking. In part, it's a case of observing what our parents and peers do and then copying them (or rebelling against them and doing the opposite). And what our parents and peers do depends on many other variables, including their education, social class and culture. For example, the incidence of smoking among teenagers in Britain is linked to the social class of their parents. The teenage daughters of unskilled or semi-skilled workers are more than twice as likely to smoke as the daughters of middle-class professionals: 50 per cent are smokers, compared with 22 per cent of those from professional homes. But social pressures are not the full story. Certain individuals are inherently more susceptible than their fellows to chemical addiction, whether it be to nicotine, alcohol or cocaine.

We saw in chapter 8 that people who are highly reactive to stressors are especially vulnerable to coronary heart disease. The roots of these variations in biological reactivity can be traced back to infancy. Babies and the young of other species respond to their environments in individually distinctive ways from birth. Some babies are consistently more reactive than others, both behaviourally and physiologically. Some babies are placid and sleep a lot, while others are irritable and constantly active. Some are easily distracted, responding warmly to attention, while others are unsociable, preferring to exist in a world of their own. These early temperamental characteristics tend to be fairly stable and consistent, so that, to a degree, they predict how that person will react to stimuli later in life.

One of the simplest ways of investigating a newborn baby's temperament is with the heel-stick test. In many countries it is standard practice to take a small blood sample from babies a few days after birth to screen them for medical disorders. This is normally done by pricking the baby's heel with a sterile lancet and squeezing out a drop or two of blood. It hurts, and babies usually react in protest, but the intensity and duration of their reaction varies considerably. Some cry loud and long; others protest little. Moreover, those who are most reactive at two or three days of age tend to remain so throughout the months that follow. Behaviourally reactive individuals also tend to be physiologically reactive: babies who display a vigorous behavioural response to the heel-stick typically show a large hormonal response, too.

Distinctive individual differences in behaviour and biological reactivity are found in the young of other species as well. Piglets exhibit consistent behavioural and physiological differences in the way they respond to their environment from a very early age. Some little piggies consistently behave in a bold, active way when placed in a novel environment, whereas other little piggies are consistently passive. These behavioural idiosyncrasies are mirrored by dissimilarities in the pigs' hormonal responses.

Early individual differences in reactivity are important because they can have repercussions for our behaviour and health later in life, although no one yet understands quite why. Scientists have found that babies who display a prolonged behavioural reaction to an injection at two months of age tend to have more illnesses when they are toddlers. Several long-term studies have found that young children whose blood pressure shows the greatest increase in response to psychological stressors are more likely to have hypertension when they are adults.

Certain forms of early experience can have a long-term impact on behaviour and immune function. Experiments conducted in the 1960s by George Solomon, Seymour Levine and colleagues found that if young rats were regularly handled by humans from birth until weaning, their immune responses in later life were distinctively different. Rats that had been handled when young produced bigger antibody responses when immunized as adults. Other researchers found that if rodents were stimulated in various ways when young, their susceptibility to certain forms of cancer in adulthood was altered. Their early experiences had affected their immune function and health in adulthood.

Stability and change feature in the equation, too. We might ask whether the traits of hostility and aggressiveness that form the core of the Type A coronary-prone behaviour pattern are modifiable or, as these words from *The Mikado* imply, deep-seated and immutable:

> I can trace my ancestry back to a protoplasmal primordial atomic globule. Consequently, my family pride is something inconceivable. I can't help it. I was born sneering.

The fact that numerous behavioural and physiological traits manifest themselves early in life and remain consistent over time does not necessarily mean that they are fixed and unchangeable. We have seen that the coronary-prone Type A behaviour pattern can be modified in adulthood, with beneficial consequences for health. Similarly, experience and training can increase our resistance to psychological stressors. We must try to escape from the Freudian notion that the icy hand of early experience will reach out and grip us in later life, no matter what we do. We are all capable of change. As Hal announces to a shocked Falstaff in Shakespeare's *Henry IV Part 2*:

> Presume not that I am the thing I was;
> For God doth know, so shall the world perceive,
> That I have turn'd away my former self...

MOTHERS AND OFFSPRING

The mother–offspring relationship is pivotal for all young mammals, including humans.[3] Without physical protection, nutrition and warmth we would all die soon after birth. Without love, stimulation and a degree of stability our mental, emotional and social faculties would be stunted.

The mother also influences her infant's immune system. Babies receive antibodies from their mothers' milk, and intimate contact between the two results in the transfer of bacteria and other antigens which stimulate the offspring's immune system. These are just two of the many reasons why breast-fed babies are, on average, physically and psychologically healthier than bottle-fed babies, both then and later. Breast really is best.

Given the central importance of the early mother–offspring relationship for physical and mental development, it is not surprising that disturbances in this relationship can have widespread repercussions for the infant, a number of which are felt much later in life. Research has established that when baby mice are separated from their mother for a few hours each day during their first two weeks of life, their adult immune response is significantly impaired. When they grow up, the separated mice produce fewer antibodies in response to immunological challenges and have less responsive lymphocytes.

Detailed long-term studies by Christopher Coe and colleagues at the University of Wisconsin have demonstrated that disturbances in early rearing conditions can have lasting effects on the immune function of monkeys, as well as perturbing their behavioural and emotional development. If rhesus monkeys are reared in nurseries, apart from their mothers, their immune systems are measurably altered when they become adults. One way in which this manifests itself is a lower level of natural killer cell activity. Monkeys that have been temporarily separated from their mothers during infancy, even for a week or two, have markedly less responsive lymphocytes when tested as mature adults years later.

We saw earlier that those with Type C personalities have a greater statistical risk of getting cancer. There is speculative evidence that Type C traits might arise in certain individuals because of disturbances in their parent–offspring relationships in infancy.

Studies have indicated that a lack of emotional closeness to one or both parents during early childhood might increase the risk of developing cancer later in life, although the evidence is far from conclusive. Cancer patients consistently report having had colder, more distant relationships with their parents than people suffering from other serious diseases. It is often unclear how objective these recollections are; they could well be distorted by current perceptions. Nevertheless, it is plausible that children who have unloving or insecure relationships with their parents might grow up exhibiting Type C traits – repressing their emotions and complying with others' wishes – as a defensive reaction to their lack of a secure emotional bond.

There is good evidence that an infant's physiological responses are affected by the security of its relationship with its mother. By one year of age, infants vary considerably in the security of their emotional attachment to their mothers. Insecurely attached infants tend to exhibit less behavioural distress than securely attached infants when temporarily separated from their mothers. Despite this outward calm, however, the physiological measurements reveal that an emotional reaction is taking place. Insecurely attached infants show a marked biological response to separation, including increases in pulse rates and cortisol levels. In effect, these infants are suppressing their outward display of emotion. But whether it is the unsatisfactory parent–child relationship or the consequential Type C personality that modifies the eventual risk of cancer remains to be seen.

Evolution

> They are in you and in me; they created us, body and mind; and their preservation is the ultimate rationale for our existence ... they go by the name of genes, and we are their survival machines.
>
> Richard Dawkins, *The Selfish Gene* (1976)

> [medical] advances would be even more rapid if medical professionals were as attuned to Darwin as they have been to Pasteur.
>
> G. C. Williams & R. M. Nesse, *Quarterly Review of Biology* (1991)

The theory of evolution by means of natural selection was first elucidated by Charles Darwin and revealed to the world in November

1859 with the publication of *The Origin of Species*, one of the most important books in human history. Ever since then, evolutionary theory has formed the conceptual bedrock of the life sciences.

To squeeze a mass of elegant science into a very small nutshell, living organisms have evolved by means of an extraordinarily powerful process called natural selection, which operates over very long timescales. This process entails the differential survival of genes which are good at replicating and getting themselves passed on to future generations. In practical terms, this boils down to the selection of genes which improve the individual organism's ability to survive and reproduce in its normal environment.

To encapsulate the sheer beauty of evolution by natural selection even more succinctly, I can do no better than quote Richard Dawkins, that modern master of Darwinian theory and most lucid of science writers:

> The world became full of organisms that have what it takes to become ancestors.

Evolution by natural selection has no grand design or ultimate goal. It does not strive to produce perfection or exquisite complexity, although exquisite complexity is often an outcome. There is no cosmic hand on the tiller, steering evolution towards a specific end result. In particular, natural selection does not select genes merely because they improve the health or happiness of individuals. All that matters as far as natural selection is concerned is the individual's ability to pass its genes on to future generations, one way or another. As Charles Darwin put it, 'perfection consists in being able to reproduce'. A gene that renders those who inherit it prone to depression or causes crippling illness in old age can still flourish and replicate, provided its net effect is to increase its contribution to future generations.

Despite the enormous conceptual power of Darwinian evolutionary theory, scientists have only recently begun to apply it to the problems of medicine. So great is the potential contribution of the evolutionary perspective to medicine that some scientists have heralded the dawn of a new age of Darwinian medicine.

Two leading proponents of Darwinian medicine (or evolutionary medicine, as some prefer to call it) are George Williams, one of the twentieth century's foremost evolutionary biologists, and Randolph Nesse, a psychiatrist and physician. In their fascinating book *Evolution and Healing*, Nesse and Williams have argued persuasively that the evolutionary perspective can offer medical science many fresh insights. Darwinian medicine is still in its infancy – in fact it is positively embryonic – but it undoubtedly has great potential to add to our understanding of disease.

A fundamental question – but one that is rarely asked – is why our bodies apparently have so many design flaws which make us unnecessarily susceptible to illness. If natural selection is sufficiently powerful to produce such exquisite pieces of biological machinery as the immune system or the human brain, then how come we still get sick all the time? Natural selection can hone organisms to near perfection, so why do millions die from heart disease, cancer and AIDS, or suffer from depression, schizophrenia, tuberculosis or parasitic infections? As George Orwell remarked in his essay *How the Poor Die*:

> People talk about the horrors of war, but what weapon has a man invented that even approaches in cruelty some of the commoner diseases? 'Natural' death, almost by definition, means something slow, smelly and painful.

Deleterious genes which make people especially susceptible to life-threatening diseases such as cancer or coronary heart disease are passed down through the generations, even though we might expect these genes to have been eliminated by natural selection thousands of years ago. An even bigger puzzle is why our bodies are constructed in such a way that psychological stress can increase our susceptibility to disease.

We shall consider several possible explanations for the fact that humans continue to be riddled with diseases. To begin with, a number of the apparent defects in our bodies turn out not to be defects after all, but biological defence mechanisms. We notice their unpleasant manifestations while overlooking their less obvious biological benefits.

We must remember, too, that the environment in which the majority of humans now live is, in evolutionary terms, very strange and very new. Our bodies evolved over millions of years to cope with a pre-industrial world. In the twinkling of an evolutionary eye, those same bodies were catapulted into the radically altered environment of modern industrialized societies. The human body and brain have not had enough time to evolve over the few thousand years in which civilization has grown – a mere trice when measured against the enormous timescale of evolution. There are aspects of our biology and psychology that are consequently ill-suited to our current environment.

A third important point is that we humans are locked in an evolutionary arms race with the bacteria, viruses and other organisms that make us ill. These disease-causing organisms (or pathogens) are subject to the same pressures of natural selection as we are. As the human species evolves ever more elaborate biological defence mechanisms to protect us from pathogens, so the pathogens are themselves under intense selection pressure to circumvent those defences, and so on, in an ever-escalating spiral. Sometimes the pathogens are ahead of us in this evolutionary arms race. At the moment, for example, the HIV retrovirus is way ahead of us.

Natural selection cannot start afresh with a new basic design for the human body every time we encounter a new medical challenge. Our bodies have evolved over millions of years by the gradual modification of pre-existing forms. Unlike an engineer designing a machine, natural selection cannot simply scrap the old model and start again. It must work by modifying the existing design. Moreover, each biological structure or characteristic has numerous ramifications, both good and bad. Natural selection acts to produce the optimal balance between costs and benefits. Diseases that have a genetic basis might persist because the 'bad' genes also have beneficial effects which outweigh their disadvantages (at least, in certain environments). Perfection cannot be expected in a changing world. We shall now look at these evolutionary ideas in greater depth.

SICK BY DESIGN

When is a defect not a defect? Many medical phenomena are regarded as solely pathological; we notice their unpleasant manifestations and give little thought to their possible benefits. Scientists and doctors are keenly interested in *how* they happen but not *why* they happen. Yet some medical phenomena only start to make sense when viewed from an evolutionary perspective.

A biologist who has been educated to think in evolutionary terms is instinctively sceptical of medical explanations that dismiss an important aspect of human nature as a gross defect in design or a pathological breakdown in the body's normal functioning. Both of these things happen and not even the most rabid Darwinian would suggest that natural selection creates perfection. Nonetheless, the evolutionary perspective is starting to reveal that some of the things which happen when we are ill should not be dismissed as annoying symptoms of disease. They are in fact biological defence mechanisms which have evolved to protect us.

One obvious and uncontentious example of such a defence mechanism is a cough. When we have an infection that irritates our respiratory tract we cough involuntarily. In doing so we expel the nasty gunge that might otherwise damage our lungs and spread further infection. This simple biological defence mechanism has presumably evolved during the history of our species because ancestral humans who coughed when they had a respiratory infection were more likely to survive and reproduce than individuals who did not cough. Other mammals cough as well, probably for the same reason. If you have a serious respiratory infection it is normal – and usually beneficial – for you to cough. Suppressing an irritating cough by taking medicines can have its drawbacks.

Fever is another example of a biological defence mechanism that is often regarded as merely a disagreeable symptom of disease. A raised body temperature is a standard biological response to bacterial and viral infection, and a good indicator of illness. Fever is commonly accompanied by changes in mental state and behaviour, including general malaise, mild depression, lethargy and loss of appetite. We feel like curling up in a dark corner and waiting to get better.

In ancient Greece, fever was viewed as an adaptive response to infection. Hippocrates came to the correct conclusion (albeit for dubious reasons) that a moderate fever assists the patient to overcome infection. This perception changed radically in later centuries. Fevers are unpleasant and therefore considered undesirable. Drugs that could make the symptoms disappear were welcomed enthusiastically. By the end of the nineteenth century the use of fever-reducing (antipyretic) drugs had become commonplace. Nowadays when we have a mild fever we reach for the aspirin or paracetamol, but the Darwinian perspective suggests that Hippocrates was nearer the mark in his attitude towards fever.

There are sound reasons for reviewing our outlook on fever. It is a universal biological response to infection in all vertebrates, from fishes to mammals. Animals are equipped with specific biological mechanisms for generating just the right amount of fever in response to infection. As a defence mechanism, it is biologically costly: animals expend considerably more energy when their body temperature is raised. Evolutionary logic implies that fever should have biological benefits to offset its obvious costs. Fever, malaise and the accompanying psychological changes are designed into our bodies. But what are they for? What is their biological function?

The current consensus is that natural selection has favoured the evolution of fever because it helps those who are sick to combat infection. Fever is designed to make life harder for the offending bacteria or viruses. Like all organisms, bacteria and viruses work best at a particular temperature. Alter that temperature and they cease to thrive.

A corollary of this evolutionary explanation for fever is that taking drugs to 'cure' fever might do more harm than good. There is evidence to support this. Experiments have shown that in certain circumstances suppressing a fever can actually impede recovery from infection. In one experiment, chickenpox sufferers were either treated with paracetamol, to lower their body temperature, or were given an inert placebo. (To avoid unconscious bias, the subjects were not told which.) Those who received paracetamol took on average a day longer to recover from chickenpox than those who were left to their own devices. 'Curing' the fever delayed their recovery.

The same thing can happen when we take aspirin or paracetamol to relieve the feverish symptoms of a common cold. Australian researchers at the University of Adelaide deliberately infected healthy volunteers with rhinoviruses to give them a cold and then treated them with a standard medicament to relieve their symptoms.[4] The tests showed that, far from bringing relief, aspirin and paracetamol made clinical symptoms worse and lowered the volunteers' antibody responses to infection; those subjects who took an inert placebo fared better.

Once again, the message in the data is that suppressing the body's natural reactions to infection can be a mixed blessing. Fever-reducing drugs are undoubtedly beneficial in many cases and can be vital when a high fever threatens the patient's health. But their indiscriminate use to suppress the mild fevers that accompany common infections might be misguided. Indeed, there might be circumstances in which it would be beneficial to take drugs that *raise* our body temperature.

Evolutionary principles have similarly cast new light on morning sickness. Once again, a medical phenomenon that has traditionally been regarded as an undesirable symptom has now been re-cast as a biological defence mechanism with distinct benefits.

The term morning sickness (or, more accurately, pregnancy sickness[5]) refers to the nausea, occasional vomiting and accompanying taste aversions which often plague women during early pregnancy. Like coughing and fever, it is a universal response and affects nearly all pregnant women to some degree. It seems to occur in populations throughout the world, including pre-industrial societies.

During the first three months of pregnancy around 55 per cent of women experience occasional vomiting and more than 75 per cent report feeling nauseous. Virtually all women develop mild or pronounced aversions to tastes and odours which they previously found palatable. As a result they tend to eat a narrower range of blander-tasting foods. In this sense, the misleadingly-named morning sickness affects nearly all pregnant women, including many who never experience actual vomiting. The sheer universality of morning sickness suggests that it might have hidden biological benefits to offset its obvious disadvantages, and biologist Margie Profet has proposed a detailed and well-supported theory to account for the evolution of this condition.

All plants, including the ones we eat, contain cocktails of toxic chemicals. These toxins evolved as the plants' way of protecting themselves against being eaten or parasitized. For example, those gorgeous aromas and tastes that we associate with freshly-brewed coffee denote the presence of several hundred chemical substances, many of them toxic. We, in turn, have evolved sophisticated biochemical mechanisms for detoxifying foodstuffs so that we can eat them with impunity. According to Profet's theory, morning sickness is not merely an annoying by-product of pregnancy; rather, it evolved to deter pregnant mothers from eating or drinking toxic substances that might harm the foetus. She has collated an impressive array of evidence to support her theory.

To start with, it is known that various plant toxins which are harmless to adults can still cause serious malformations in developing embryos or even trigger spontaneous abortions. Certain plant extracts have been used for centuries to terminate unwanted pregnancies.

Secondly, the time course of morning sickness runs parallel to the embryo's changing vulnerability to toxins. The condition tends to manifest itself around the third or fourth week after conception, peak during the third month, and then subside. This matches the period of maximum vulnerability for the developing embryo. The organs and limbs are mostly formed between three and eight weeks of age and it is during this period that the embryo is most vulnerable to toxins in the mother's diet. By four months of age the worst danger is past and the embryo has an overriding need for energy to fuel its growth; by this stage the morning sickness has normally diminished and the mother's feeding habits are returning to normal.

The nature of morning sickness is consistent with its putative biological function. It is typically accompanied by an aversion to certain foodstuffs, especially those with a strong or bitter taste. Hormonal changes affect the pregnant woman's sensory mechanisms and make her much more sensitive to taste and odour cues emitted by food and drink. Many of the natural toxins that are found in foodstuffs have distinctive tastes and odours that correspond reasonably well with the tastes and odours which pregnant women tend to find least attractive. Although women differ greatly in their particular likes and dislikes during pregnancy, most find they are less attracted to pungent or bitter-tasting substances such as coffee, spicy foods, herbs, strong alcoholic drinks and barbecued or fried foods, and their diet will therefore become blander than usual.

Perhaps the real proof of the pudding is the fact that women who experience pronounced morning sickness during early pregnancy are significantly less likely to suffer a miscarriage or give birth to a malformed baby than women whose morning sickness is mild or negligible. According to various estimates, pregnancies end in spontaneous abortion in approximately 1–3 per cent of women who experience severe nausea and vomiting, as compared with 4–7 per cent of women who experience little or no overt morning sickness.

As in the cases of coughs and fever, Profet's evolutionary hypothesis about morning sickness implies a counsel of caution. If the condition is indeed a biological defence mechanism, designed to protect the developing embryo, it follows that pregnant women should think twice before taking drugs to 'cure' it. Nowadays the use of anti-nausea drugs and sedatives to suppress morning sickness is regarded with much greater circumspection than hitherto, partly as a result of the thalidomide disaster. The sedative and anti-nausea drug thalidomide was withdrawn from use in the 1960s after women who had taken it gave birth to babies with deformed or missing limbs. The thalidomide disaster illustrates how vulnerable the embryo is during its first three months to chemicals that cause developmental abnormalities.[6]

Another phenomenon that has often been treated as a symptom of disease rather than a biological defence mechanism is the sequestration of iron – a form of iron 'deficiency' which accompanies some infections. The average human body contains about four grams of iron, mostly in the red blood cells. However, when we have an infection the amount of iron circulating in our blood may drop to as little as a fifth of its normal concentration. This drop in iron level happens because the body actively removes iron from circulation and sequesters it in the liver. Like fever, it happens for a reason. Bacteria need iron to grow and reproduce; depriving them of this vital resource is therefore an effective tactic in the body's battle against bacterial infection. Giving infected patients supplements of iron to correct their 'deficiency' is akin to sending Red Cross parcels to the offending bacteria.

What all these examples – coughs, fever, morning sickness and iron sequestration – illustrate is that we dismiss medical phenomena as mere symptoms at our peril. They may be manifestations of biological defence mechanisms which have evolved for a reason and which have benefits as well as costs. We would do well to understand the nature of these benefits before we take drugs to suppress them. Tolstoy memorably expressed this point in *War and Peace*:

> Our body is a machine for living. It is organized for that, it is its nature. Let life go on in it unhindered and let it defend itself, it will do more than if you paralyse it by encumbering it with remedies.

GENES FOR DISEASE

Many diseases have a genetic basis – that is, certain genes carry with them a susceptibility to particular diseases. Where hereditary diseases like cystic fibrosis and Huntington's disease are concerned, there is a reasonably straightforward relationship between the faulty gene and the disease. In a few cases, scientists have even worked out the molecular mechanisms which lead from the faulty gene to the disease symptoms. Cystic fibrosis, for example, arises because a faulty gene on chromosome number 7 prevents the production of a protein which is essential for the normal functioning of cell membranes. Individuals who lack this protein produce thick mucus which blocks their airways, pancreas and intestinal glands, with devastating consequences for their health. But in the majority of cases, including those genes which predispose the inheritor to heart disease or cancer, the causal connections between the genes and the disease are inextricably convoluted.

An obvious question is why natural selection has allowed these 'bad' genes to persist in the gene pool. Defective genes can result in serious disease or death, so why have they not been eliminated from the gene pool during the course of evolution? Surely organisms which lacked the 'bad' genes would be better at surviving and reproducing than those who possessed them? Not necessarily.

Evolutionary biologists have pointed out various ways in which genes that predispose the carrier to disease could continue to be passed on through the generations. First of all, a gene that increases the risk of developing a disease can be maintained by natural selection if it has additional, beneficial consequences which outweigh its disadvantages. Any one gene will exert a range of influences on the structure and functioning of an organism, some of which may be beneficial and some of which may be deleterious. It is the net effect of these advantages and disadvantages on the organism's survival and reproduction that decides whether the gene is favoured by natural selection. If the selective advantages outweigh the disadvantages then the gene will continue to be passed on to subsequent generations.

The traditional textbook example of this evolutionary balancing act is the gene responsible for the hereditary blood disorder sickle-cell disease. The disease occurs when an individual inherits two copies of the sickle-cell gene (one from each parent). In most cases, however, a single copy of the sickle-cell gene is inherited from one parent, making the recipient a carrier of the disease. Carriers normally show no symptoms and tend to be more resistant to malaria than people who lack the sickle-cell gene. In tropical regions where malaria is rife this resistance is (or certainly was) a significant advantage. The sickle-cell gene has therefore lived on because it confers biological benefits, at least in some environments. As you would predict, the sickle-cell gene is more prevalent in malarial regions.

A more common way in which disease-causing genes persist in the gene pool is when their bad effects do not manifest themselves until late in life, after the organism which carries the gene has finished reproducing. Natural selection operates on genes which have a bearing on the organism's success at surviving and reproducing, and is therefore relatively blind to the effects of defective genes which do not express themselves until the reproductive years are long past. The selection pressure to eliminate the gene will be weak if the gene has no tangible effects until late in middle- or old-age.

As far as natural selection is concerned, what happens to your health after you are too old to reproduce or care for your offspring is largely irrelevant. (Largely, but not completely irrelevant; there is still a degree of selective pressure on the grandparents' genes because looking after grandchildren affects *their* chances of surviving and reproducing.) This explains why inherited diseases such as Huntington's disease do not emerge until middle age. A defective gene that caused organisms to become seriously ill and die before they were old enough to reproduce would never give carriers the chance to pass it on to future generations.

If the 'bad' gene also does good things early in development then its later disease-causing effects will be even less important in the process of natural selection. Genes that cause bits of you to malfunction when you are old and crumbly can replicate themselves, provided they help you to survive and reproduce in your youth. Evolutionary biologists believe that many of the genes involved in the degenerative diseases of senescence have persisted in the gene pool because they do not normally express themselves until after the age at which most individuals have finished reproducing. (And remember that we are talking about Stone Age hunter-gatherers here. For our evolutionary forebears, forty would have been old.)

One intriguing implication of this evolutionary logic is that certain genes which underlie diseases of senescence, such as Alzheimer's, osteoarthritis and various cancers, might also have important benefits earlier in life. Eliminating these 'bad' genes through new gene therapy techniques could therefore have unforeseen consequences. Once again, the counsel of caution from evolutionary biology is that we should be careful about tinkering with our bodies until we really understand what we are doing.

DISEASES OF MODERN LIFE

Our bodies may seem prone to defects because our biological nature is no longer well matched to our environment. Our bodies and minds have been designed by natural selection to work optimally in the environment in which the human species has evolved – a place biologists sometimes refer to as the environment of evolutionary adaptedness. The trouble is that this optimal environment bears little relation to the environment in which most of us now live.

But before we move on let me reiterate a crucial point, just in case there is scope for misunderstanding about my use of the word 'design'. Natural selection has no ultimate goal or purpose or grand design. It produces organisms that look as though they were designed by a clever engineer, but they were not. We are the way we are because our ancestors were successful at being ancestors; because our genes were good at replicating themselves. We are, to quote Democritus, 'the fruit of chance and necessity'. It is useful shorthand to talk in terms of our bodies being designed, but it is only shorthand.

The human body and mind are simply not built to cope with life as an office worker in London, or as a farmer in Moose Jaw for that matter. The environment that natural selection has spent hundreds of thousands of years honing us for is essentially that of a Stone Age hunter-gatherer. Our bodies and minds have evolved to work optimally in one world, but we have very recently and very suddenly (in evolutionary terms) created for ourselves a radically different world. The resulting mismatch between our biological nature and our current environment has inevitably generated problems.

In terms of human evolutionary history, the physical and social environments of modern industrialized civilizations are extremely novel. To realize just how novel, it is necessary to grasp the timescales involved.

The first recognizable forms of organic life appeared on earth around 4,000,000,000 years ago, slopping around in the primordial soup. The first animals with backbones appeared around 500,000,000 years ago. Our earliest hominid ('ape-man') ancestors started diverging from the apes about 7,000,000 years ago and they began to walk upright on two feet 3–4,000,000 years ago. Some time after this, probably around 2,500,000 years ago, they started to make the first primitive stone tools.

Homo sapiens sapiens, the particular primate species to which you and I belong, did not appear on the scene until much more recently. Analysis of DNA sequences indicates that the first true humans were walking on earth only 200,000 years ago. The oldest human fossils are a trifling 100–150,000 years old. The whole history of the human species accounts for only about 3 per cent of the time since our first hominid ancestors appeared on earth.

The human timescale shrinks to a diminutive pin-prick when we consider the origins of civilization. Human civilization is less than

10,000 years old and has therefore occupied less than 5 per cent of our species' history. The earliest forms of agriculture were invented around 10,000 years ago; the first cities were built less than 7,000 years ago; and writing appeared about 6,000 years ago. The wheel was invented in the Middle East in about 3500 BC, or less than 300 human generations ago. Jesus of Nazareth was born barely 100 generations ago and industrialized civilization is barely a few generations old.[7] Modern industrialized civilization has existed for less than one thousandth of the evolutionary history of our species.

Imagine the whole 4,000,000,000-year history of life on earth compressed into a single year. The first primitive hominids materialize during the morning of 31 December, *Homo sapiens sapiens* appears barely half an hour before midnight, and the first primitive civilizations pop into existence less than two minutes before the bells start to ring in the New Year. In evolutionary terms, the whole of human civilization has existed for the merest fraction of an instant.

Genetically, you and I are almost identical to the humans who were around at the dawn of civilization. Scientists can estimate the genetic difference between us and our hunter-gatherer forebears, based on rates of change in DNA sequences. Spontaneous mutations alter the DNA in the nuclei of our cells at an average rate of about 0.5 per cent every 1,000,000 years. This means that in the 10,000 years or so since the earliest human civilizations started to emerge, our genetic make-up has probably changed by only about 0.005 per cent, or one part in twenty thousand. As George Williams and Randolph Nesse have pointed out, a hunter-gatherer baby born thousands of years before the wheel was invented would be biologically capable of growing up to become a lawyer or an accountant (poor thing).

Natural selection has had nothing like enough time to equip us with the biological and psychological machinery needed to handle the problems and temptations of the modern world. Many of the so-called diseases of civilization which now plague us, such as heart disease and cancer, result partly from the mismatch between our biology and our current environment.

As far as scientists can deduce from the fragments of archaeological and biological evidence, we humans spent the great bulk of our evolutionary history as hunter-gatherers. We lived in small tribal groups, foraging for whatever plant foods happened to be available and hunting the occasional animal. Food supplies were often unpredictable in our hunter-gatherer environment of evolutionary adaptedness. High-energy fats and sugars were particularly scarce and valuable resources. It therefore made good biological sense for humans to perceive fatty or sweet foods as deeply desirable and to eat lots of them whenever they were available.

This trait was extremely adaptive on the grasslands of Africa 150,000 years ago, but nowadays it is much less of a boon to wealthy, over-fed Europeans and Americans. The bulging bellies, wobbling buttocks, clogged arteries and cancers of the colon that plague most industrialized societies are a dreadful testament to the mismatch between our biological heritage and the new environment it has helped us to create. The psychological characteristics that might enable people to resist the lure of greasy hamburgers, cigarettes, indolence and fast cars simply have not had time to evolve. Meanwhile we have to make do as best we can with conscious decisions based on rational thought. Some hope.

Our physical environment is not the only thing to have changed radically within the recent evolutionary past. Our social environment was also very different. The best available evidence indicates that our hunter-gatherer forebears lived in stable, close-knit social groups with between twenty and a hundred members. Most of the individuals in the social group would have been genetic kin and all would have known each other fairly well. Having long-term, mutually supportive social relationships with several other individuals within a stable social group was probably a standard feature of life throughout much of human history.

Contrast this with the social environment of solitary urban dwellers now. Surrounded by countless thousands of people, they may know hardly anyone. And, as we saw in chapter 6, our social environment matters a great deal for our mental and physical health.

Patterns of disease and mortality have also changed profoundly within the recent past, thanks to improvements in living conditions and enormous advances in medicine. In industrialized nations with organized health care and social welfare systems it is now relatively rare for people to die from malnutrition, bacterial infections or the effects of parasites. Yet these were the great killers of days gone by, and our bodies' defence mechanisms evolved accordingly. What history has not programmed our bodies for is to defend against the effects of smoking, prolonged inactivity or constant overeating.

Humankind's recent departure from its environment of evolutionary adaptedness might help to explain historical changes in the incidence of certain cancers. A number of scientists have argued that the comparatively high rate of breast cancer and ovarian cancer among women in industrialized societies stems partly from long-term changes in human reproductive patterns. Epidemiological evidence suggests that never having any babies, having few babies, or having your first baby when you are relatively old, are all factors which increase a woman's risks of developing what are known as reproductive cancers (i.e., cancer of the breast or uterus). Since the Second World War, however, the trend in industrialized societies has been for women to have fewer babies, later in life.

Our current reproductive pattern is very different from what it was in our environment of evolutionary adaptedness. Some insight into past reproductive patterns can be gained by looking at the few surviving hunter-gatherer societies such as the !Kung San of Africa, whose lifestyles are almost certainly closer to those of our evolutionary forebears than yours or mine. On this basis a woman's 'natural' reproductive pattern – that is, the one for which her genes have equipped her – is probably to have five or more babies starting in her late teens, and to suckle each baby for three or four years (during which time she would not ovulate). A radical change from this reproductive pattern, though eminently desirable for many other reasons, could have hidden biological costs.

The world has moved on, and I am definitely not implying that we must all return to a Stone Age lifestyle if we wish to be healthy and sane. The human body and mind are immensely adaptable, allowing us to find the means to cope with most circumstances, no matter how bizarre.

EVOLUTIONARY ARMS RACES

Another reason why all is not for the best in the best of all possible worlds is that disease-causing organisms are frequently ahead of us in the great evolutionary arms race.

We are not the only species on whom natural selection has been working its magic. The various species of bacteria, viruses, parasites and fungi which would, given half a chance, set up home inside your cosy and nutritious organs have also been subject to its influence. Natural selection has equipped us with a formidable array of sophisticated defence mechanisms, including the immune system, to protect us against these pathogens. If our biological defence mechanisms were not highly effective then I would not be here writing about them and you would not be there reading about them. But the enemy have been similarly equipped to fight the battle for survival; natural selection has armed them with the ability to circumvent human defence mechanisms.

The evolutionary relationship between a host organism (such as you or me) and invading pathogens (bacteria, viruses or parasites) is in many respects analogous to the arms race that occurred between the superpowers during the dark days of the Cold War. The two sides were locked into an endlessly escalating spiral of threat and counter-threat, new weapons and countermeasures, leading to ever more elaborate and costly weaponry. One side devised a smart new radar to detect incoming missiles, so the opposition deployed even smarter missiles to circumvent the new radar, and so on *ad Star Wars*.

A comparable process occurs in the evolutionary arms race between host and pathogen. Both the host and the pathogen are evolving under the pressures of natural selection. Genes that help organisms to survive and reproduce are selected. As the host species evolves new tricks for defending itself against pathogens, so those pathogens are under intense selection pressure to avoid being annihilated by the host's defences. Sometimes the host is ahead in the arms race and sometimes the pathogen is ahead.

Natural selection has equipped bacteria, viruses and parasites with an assortment of sophisticated tricks for outwitting our defences. For instance, the parasitic micro-organism which causes African sleeping sickness (trypanosomiasis) has evolved the ploy of continuously changing its biochemical appearance, so that our immune system is incapable of locking onto it and generating specific antibodies to destroy it. Streptococcal bacteria have evolved a biochemical coat which mimics human cells, making it difficult for our immune system to recognize the bacteria as antigens. This immunological mimicry can cause major problems. So similar are these bacteria to human cells in their outward appearance that the immune system can be provoked into attacking its own body tissue by mistake. Hence, streptococcal infections may occasionally result in rheumatic fever, a disease in which autoimmune reactions cause severe damage to the heart and joints. Another ploy used by some pathogens is to hide inside the cells of the immune system itself. HIV exploits this clever ruse, making itself an especially difficult target for our immune defences.

We must expect to be on the losing side in the evolutionary arms race at least some of the time. Bacteria and viruses have a huge advantage over us because they are minute, multitudinous and prolific reproducers. No matter how fastidious you are about your personal hygiene, there are more bacteria inhabiting your body than there are human beings on the planet. And those bacteria are reproducing several hundred thousand times faster than you could manage. Humans take twenty or thirty years to produce a new generation. Bacteria can do it in thirty minutes. Under the right conditions, they can whizz through tens of generations in the course of one day, reaping the same evolutionary advantages that would take us hundreds of years to achieve. Evolution by natural selection is going on inside your body right now.

The HIV retrovirus is particularly adept at the evolutionary game and can modify its genetic structure about one million times faster than its human hosts. This enormously rapid rate of genetic change presents huge problems for the human immune system – and for the designers of potential AIDS vaccines. Before the virus's weak spot can be identified, it has changed.

One of the more worrying consequences of recent selection pressures has been the evolution of new strains of bacteria that are resistant to antibiotics. Since the end of the Second World War, when antibiotics first came into widespread use, many species of bacteria have evolved effective defences against antibiotics. Bacteria that happened to have genes which made them resistant to antibiotics thrived and passed those genes on to subsequent generations, whereas those that lacked resistance were wiped out.

The selection pressure for resistance has been intensified by the excessive and indiscriminate use of antibiotics. The practice of routinely putting antibiotics into animal feed has done much to improve agricultural yields, but in the process it has hastened the emergence of antibiotic-resistant strains of Salmonella which now threaten human health. Moreover, bacteria are able to swop genetic material without having to reproduce. A single bacterium can transfer a resistance gene horizontally to its neighbours, enabling resistance to antibiotics to spread very rapidly. Resistant strains of bacteria have been known to supplant susceptible strains of the same species within the course of a few weeks. Fifty years ago there were no penicillin-resistant strains of staphylococci (the bacteria responsible for many post-operative infections). Now, nearly all staphylococci are resistant to penicillin and a range of other antibiotics besides. Much the same has happened with the drugs used to combat malaria. Natural selection has produced resistant strains of the protozoan parasites which cause malaria, rendering one after another of the standard anti-malarial drugs ineffective.

THE FUNCTIONS OF UNPLEASANTNESS

The mind–body phenomena we have explored in this book present some of the greatest challenges for evolutionary theory. The Darwinian perspective should be able to make sense of psychoneuroimmunology.

A deceptively simple question is why (as opposed to how) psychological and emotional factors make us ill. Of what possible adaptive value are those elaborate neural and chemical mechanisms which enable psychological stress to impair the immune system and hence increase our susceptibility to disease? And why do so many people spend so much of their lives feeling anxious, depressed or just fed up?

Evolution by natural selection is, to repeat the familiar refrain, all about reproductive success and the differential survival of genes. It has nothing to do with contentment or physical wellbeing, except in so far as they may affect an organism's ability to survive and reproduce. Natural selection has no remit to make us happy or healthy all our lives. We have seen that genes can thrive in the gene pool and pass on through the generations even if they cause diseases, provided they have countervailing benefits or do not impair the carrier's ability to produce children and grandchildren. The same is true for genes that predispose us to unhappiness, depression and anxiety.

A moment's reflection will reveal that pain, nausea, fear, anxiety, sadness and most other disagreeable sensations are not wholly bad. On the contrary, they have a vital biological function. That function is to protect us from harm. People who lack the ability to feel pain, for example, are prone to injure themselves and die prematurely. Physical pain forces us to avoid harm and helps us to survive. Excruciating pain might be unpleasant, but it works – and that is all that counts as far as natural selection is concerned. As the Greek philosopher Epicurus wrote over two thousand years ago: 'The occurrence of certain bodily pains assists us in guarding against others like them.'

A similar argument applies to our universal capacity for repugnant mental and emotional states such as fear, anxiety and sadness. Anxiety is a case in point; it can prevent us from straying into danger or it can spur us on to greater efforts when we are confronted with challenges. In a dangerous world, animals who feel twitchy or anxious tend to survive longer than those which lie back in relaxed contentment. A complacent mouse would end up as a cat's dinner, whereas his twitchy, anxious cousin might survive to twitch (and procreate) another day.

Sadness, an emotion which affects all of us from time to time, also has its potential biological benefits. Sadness or mild depression is a common reaction to setbacks or losses. Some theoreticians have suggested that this emotional response might be nature's rather crude way of impressing upon us that our previous behaviour has been inappropriate. When we feel sad we are less likely to persist with what we have been doing. Giving up and trying an alternative course of action may be the best thing to do under adverse circumstances.

To an evolutionary biologist, then, negative emotions like anxiety and sadness are not merely pathological abnormalities caused by defects in our mental apparatus. On the contrary, they can be viewed as somewhat coarse protective mechanisms, designed to keep us from harm in the same way as pain, fear, revulsion or nausea keep us from harm. The biological capacity to experience such emotions is perfectly normal and, in terms of natural selection, a Good Thing. As long as it helps to increase our chances of surviving and reproducing, natural selection will continue to equip us with the mental capacity to feel bad.

Pursuing this Darwinian logic, George Williams and Randolph Nesse have made the intriguing suggestion that some people suffer from too *little* anxiety rather than too much. Being too relaxed and complacent might be risky. It is possible that those annoyingly laid-back individuals who never seem to worry about anything, and whom the rest of us envy, pay a price for their blissful serenity. Perhaps they run bigger risks of ending up in a casualty department or embroiled in a personal crisis. At last, the Achilles heel of the Type B. It would be nice to know.

The mortality statistics prove that we are nowhere near anxious enough about the commonplace dangers we face in modern industrialized societies. Driving around in cars and smoking are, in actuarial terms, two of the riskiest things that the majority of people ever do, but they are activities which most of us approach with an alarming lack of concern.

It is also interesting to speculate about the possible drawbacks of antidepressant drugs such as fluoxetine (alias Prozac) which are now in widespread use. By suppressing the unpleasant emotions that assail everyone from time to time these drugs make life more bearable. But those emotions are probably there for a reason, like coughs and fevers and morning sickness. Suppressing a person's fever when they have a viral infection can impede their recovery. Indiscriminately suppressing negative emotions may have its downside too.

A world from which all unhappiness, aggression and disagreeable emotions have been banished chemically is chillingly portrayed in *Brave New World*, Aldous Huxley's dystopian vision of the new world order in the year 632 After Ford. The unbending organization of the World State is maintained through comprehensive programmes of genetic engineering and Pavlovian conditioning, ensuring that each individual fits neatly into his or her predetermined caste role. Happiness and tranquillity are created through liberal doses of the psychotropic drug *soma*, courtesy of the state.

Soma, the universal panacea, ensures that everyone is good-natured, content and free from any disruptively original thoughts. If someone feels a tinge of disgruntlement the advice is always the same:

> What you need is a gramme of *soma* . . . All the advantages of Christianity and alcohol; none of their defects . . . One cubic centimetre cures ten gloomy sentiments.

Only one person, the Savage, is recognizably human in Huxley's deeply unappealing world, and he claims the right to be unhappy.

But back to current reality. It is all very well for evolutionary biologists to argue that a capacity for unhappiness is natural and antidepressants are the work of the devil, but what happens when these negative emotions get out of hand and lead to serious medical problems? Millions of lives are ruined by severe depression or anxiety disorders which render their sufferers incapable of functioning normally. In most industrialized societies approximately 2–3 per cent of the population is suffering from a serious depressive illness at any one time. Depression drives thousands of people to kill themselves each year, and for young adults suicide is now the second most common cause of death (the commonest being vehicle accidents).

The sheer universality of depression – or, at least, the capacity to become depressed – suggests that its underlying biological mechanisms are a basic feature of human nature. So why has natural selection equipped us with the capacity for something as disabling as depression? How come we are designed like this?

Some forms of clinical depression have a genetic basis, so whichever genes predispose individuals to depression might have other, offsetting advantages. One highly speculative suggestion has been that genes which make a person prone to depression (a Bad Thing) also make them intellectually creative (a Good Thing). In certain environments the benefits of being creative, and therefore better able to solve novel problems, outweigh the disadvantages of being cast into gloom. Because of this compensatory advantage, the gene or genes underlying the disposition to depression are maintained in the gene pool by natural selection. There is little solid evidence as yet to support this theory, though it does fit in with the informal impression that creative types – artists, writers, scientists and musicians – have suffered from more than their fair share of depression.

If this seems like clutching at straws then evolutionary biologists have offered another theory to account for our propensity to get depressed. It goes like this. Imagine you are an engineer designing a protective alarm system – a burglar alarm or smoke detector, for instance, or even a missile early-warning system. A fundamental design problem with all such devices is how sensitive they should be. Set the sensitivity too high and you will get lots of false alarms because the system responds to irrelevant disturbances. Your over-sensitive burglar alarm will be triggered by every breeze or passing fly. Set the sensitivity too low and you risk missing genuine threats. Your under-sensitive burglar alarm will not annoy you with false alarms but it could fail in its primary purpose of detecting the presence of an intruder.

Choosing the optimal sensitivity for a protective mechanism is a matter of balancing the risk of false alarms against the risk of failing to respond to genuine threats. It is usually regarded as preferable to have an over-sensitive burglar alarm or smoke detector, which occasionally disturbs you with false alarms, than an under-sensitive one which allows a burglar to steal your possessions or lets you fry in your bed when your house catches fire.

Our emotional systems, so the hypothesis goes, are analogous to burglar alarms or smoke detectors in this respect. Being anxious or sad for no good reason – an emotional 'false alarm' – is disagreeable and inconvenient. But failing to respond adequately to genuine hazards could prove fatal. In harsh evolutionary terms it is therefore better to experience these emotional false alarms than to blunder into a life-threatening situation because you were not anxious or unhappy enough. Occasionally, the protective mechanism is set too far in the direction of over-sensitivity, resulting in an anxiety disorder. This is a psychological equivalent of an autoimmune disease – a pathological state brought about by an over-zealous defence mechanism, or mental 'friendly fire'.

WHY DOES STRESS MAKE US ILL?

We turn now to an even thornier evolutionary puzzle. Why has natural selection equipped us with biological and psychological mechanisms which are capable of increasing our susceptibility to diseases and making us more likely to die prematurely? Is death merely nature's way of telling us to slow down, as an anonymous wit once suggested?

The majority of processes that occur inside us during a stress response are clearly beneficial. They prepare us to cope with threats or demands. We breathe faster, see better, pump more energy to our muscles, and so on. Brief stressors often enhance certain aspects of immune function as well. But other aspects of the stress response are less obviously adaptive. One particular puzzle is why prolonged stress stimulates our adrenal glands to pump out glucocorticoid hormones which have the effect (among other things) of suppressing the immune system and thereby increasing our vulnerability to disease.

The traditional explanation is that stress-related immune suppression is merely an unfortunate by-product of other biological processes, with no adaptive function or rationale of its own. However, this is a deeply unsatisfactory explanation. As we have seen, stress-induced changes in immune function are mediated by complex chemical and neural mechanisms which appear to have evolved specifically for this purpose.

Before dismissing immune suppression as simply a pathological defect, we should first consider whether it might have any biological functions in its own right. Perhaps stress-induced immune suppression has evolved because it has biological benefits, in the same way as coughs, fevers and morning sickness do. But what could be beneficial about suppressing the immune system?

Once again, the smoke detector analogy comes to our rescue.

The immune system, so the argument goes, is like any protective mechanism. There is the familiar problem of striking an optimal balance between responding too readily or not readily enough. This is a difficult balancing act and the immune system occasionally gets it wrong. When it fails to respond adequately to bacteria, viruses or cancer cells, we may get a fatal disease. When, on the other hand, the immune system is too vigorous or indiscriminate in its response, we may instead develop an autoimmune disease. So far, so good.

Many stressors initially stimulate an increase in immune function – the common sense prediction of what should happen when an organism is threatened with physical danger. Should the stressor persist, however, the revved-up immune system must be damped down to prevent an autoimmune disorder occurring. The immune-suppressing glucocorticoid hormones, such as cortisol, play an important role in this delicate immunological balancing act, repressing the immune system to prevent it from over-reacting and damaging the body's own tissues. The immune-suppressing hormones are there, it is argued, to prevent immunological 'friendly fire'.

The story is obviously more complicated than this. It cannot simply be a matter of control mechanisms reining-in an over-stimulated immune system, because prolonged or severe stress causes immune function to drop *below* normal levels. Indeed, most of the research on stress has concentrated on these immune-suppressive effects. Why might it be beneficial to suppress our immune functions below their normal levels? One theory is that certain viruses alter the surface markers of their host's cells, thereby making those cells appear foreign, or 'non-self'. If the host's immune system launches a massive, unrestrained attack against these virus-infected 'foreign' cells, there is a danger that unchanged cells may be damaged in the process. The end result would be an autoimmune disease. The best way of preventing this may be to suppress the immune response.

An alternative explanation rests upon the evolutionary idea we considered earlier when trying to account for diseases of civilization, and blames the mismatch between our biological makeup and our current environment. Stress-induced immune suppression might be an example of a biological phenomenon which is normally adaptive, but which is triggered inappropriately in our present environment.

According to this theory, the biological mechanisms which damp down the immune system in response to stress were designed by natural selection to cope with a different environment from the one we now inhabit. Specifically, they evolved to cope with relatively brief stressors such as the proverbial sabre-toothed tiger (or, more probably, the club-wielding bruiser from the rival hunter-gatherer group across the valley). The control mechanisms did not evolve to cope with stressors which routinely persist for weeks, months or even years – stressors which have become the hallmark of the twentieth century.

The theory is a plausible one and definitely an improvement on the notion that stress-related immune suppression is solely an incidental by-product of other processes. But it is far from perfect. For a start, it assumes that our hunter-gatherer ancestors were seldom exposed to prolonged stressors and rarely suffered the damaging effects of chronic stress. This assumption is questionable. True, the hazards of Stone Age life would have been, on the whole, short-lived. You fought or ran away or were killed. But what about the stress generated by other people? As we have seen, social relationships can have profound effects on immune function and health. Our ancestors lived in tight-knit groups, providing tremendous scope for prolonged, inescapable social stress. If other socially-living primates are anything to go by, our ancestors would often have been pretty foul to each other. Anyone who has observed groups of rhesus monkeys or chimpanzees in the wild will know just how foul.

I am not convinced that prolonged psychological stress is a uniquely modern phenomenon. We shall probably never know for certain and any evidence is bound to be tenuous (even if someone invents a new discipline of palaeopsychoneuroimmunology). If, as I suspect, our hunter-gatherer ancestors also suffered from chronic social stress and stress-induced immune suppression, the evolutionary mystery of why it happens remains unsolved.

Darwin's illness

No book which touches on health and evolution would be complete without a word or two about Darwin's illness. Charles Darwin, the most profoundly influential biologist who ever lived, was a sick man for much of his adult life.

From his early thirties until his death at the age of seventy-three, Darwin suffered from a chronic illness which rendered him a virtual invalid. In his youth Darwin had been the picture of health, the very model of an athletic, robust and sporting young man. As a student at Cambridge he had spent much of his time in outdoor pursuits: collecting beetles, shooting and fox-hunting.

But within a year of returning from his historic voyage of exploration on HMS *Beagle*, Darwin was stricken with illness. For most of the rest of his life he suffered from recurrent bouts of gastric pains, flatulence, nausea, fevers, palpitations, lethargy, muscular weakness, pains around the heart, ringing in the ears, visual problems, acute anxiety, depression and insomnia. At the age of thirty-three Charles Darwin retreated to Down House in Kent, where he was to stay, a partial invalid, until his death in 1882. As his son Francis Darwin later wrote:

> . . . it is . . . a principal feature of his life, that for nearly forty years he never knew one day of the health of ordinary men, and that thus his life was one long struggle against the weariness and strain of sickness.

What was wrong with Charles Darwin? The true nature of Darwin's illness was, and remains, a major bone of scholarly contention. It baffled Darwin's doctors at the time and has been a continuing source of academic controversy ever since.

In many respects, though, the most interesting aspect of Darwin's illness was not its true medical nature, but rather the implicit assumptions that people (including Darwin himself) made about his illness. Darwin's illness provides a further illustration of how mind–body dualism has led to the classification of illnesses as either 'mental' or 'physical', as though these were two intrinsically different and mutually exclusive categories. With a few honourable exceptions, scholars who have held forth on the subject of Darwin's illness fall into one of two opposed camps: the psychological and the physical.

In the blue corner are those who maintain that Darwin's illness was a form of neurotic psychological complaint. This view was especially prevalent during Darwin's lifetime, mainly because his symptoms did not correspond to any known organic disease. Darwin consulted most of the leading physicians of his day, but none of them could find any firm indications of organic disease. Darwin reluctantly concluded that he must be a hypochondriac. This was an unpalatable thought for a man of science and must have made the burden of his illness even harder to bear.

Of the many psychological explanations proposed for Darwin's illness, one favourite theory is that it arose from an unconscious emotional conflict, which had in turn been brought about by Darwin's deep discomfort at the impact his evolutionary theories would have on religious belief.

The Origin of Species, whilst not wholly incompatible with Christian doctrine, did little to bolster the concept of God as the omnipotent creator and designer of all living creatures. Victorian England was a nation of staunch religious belief. Darwin himself had drifted from conventional piety into resolute scepticism, but with the exception of a few intellectuals, most of his contemporaries were committed Christians, including his beloved wife Emma.[8] Darwin's realization that his evolutionary theories would cause an immense furore was an understandable source of anxiety for him. A variant on this theme is the theory that Darwin was emotionally traumatized by the hostile reaction his theories provoked in many quarters – including a number of eminent scientists – after *The Origin of Species* was eventually published.

Freudian theorists, too, have speculated about the psychological origins of Darwin's symptoms, asserting that he was the victim of an Oedipus complex, his organic symptoms born of an unconscious emotional conflict between sexual attraction towards his mother and fear of his intimidating father. A more plausible suggestion is that

Darwin's adult problems stemmed from the emotional trauma of losing his mother when he was still a child. (She died after a serious illness when he was only eight years old and there are indications that Darwin suffered long-term psychological repercussions from this early bereavement.) As we have seen, bereavement can generate significant mental and physical problems.

In the red corner, scornfully brushing aside the psychological theories, are those who maintain that Darwin suffered from a straightforward, but undiagnosed, physical disease. Since the 1960s there has been widespread support for the view that Darwin had Chagas' disease (otherwise known as American trypanosomiasis), a disabling parasitic infection akin to African sleeping sickness. Chagas' disease is caused by a micro-organism called *Trypanosoma cruzi* and is transmitted to humans by the faeces of bloodsucking bugs. According to this theory, Darwin contracted Chagas' disease during the *Beagle* expedition. Scholars have used Darwin's own notebooks to pinpoint the precise date on which he was infected – 26 March 1835. On this day, he recorded having been bitten by a large bloodsucking bug while he was ashore exploring in Argentina.

The Chagas' disease hypothesis is compelling. Darwin's symptoms were consistent with the variable and diffuse symptoms of Chagas' disease and he was certainly at risk during the *Beagle* expedition. After the *Beagle* arrived in South American waters he spent a fair amount of time sleeping outdoors in infested areas of countryside and the bug depicted in his diary fits the description.

The fact that Darwin's illness proved impossible to diagnose at the time is unsurprising. Chagas' disease is normally found in rural areas of South America rather than Kent, and the protozoan parasite that causes it was not identified until twenty-seven years after Darwin's death. Even nowadays, Chagas' disease is tricky to identify and is apt to be misdiagnosed as a psychiatric disorder. The disease runs a variable course and its manifestations differ greatly from case to case. In some sufferers, the gradual destruction of muscle and nerve cells may not produce symptoms until years after the initial infection. Peter Medawar has suggested that Darwin's health was further damaged by the Victorian medical treatments he received for his illness. Thanks to the efforts of his many doctors, Darwin may have had to cope with the effects of mercury and arsenic poisoning on top of his primary disease.

Yet the case for Chagas' disease is not conclusive. Darwin's symptoms – notably his palpitations and gastric problems – started troubling him as early as 1831. This was before he set sail on the *Beagle* and before he could possibly have contracted Chagas' disease. Moreover, during his final decade the symptoms eased considerably, resulting in a marked improvement in Darwin's health. Critics of the Chagas' disease hypothesis point to the compelling evidence that psychological factors played an important role in provoking Darwin's symptoms. As Darwin himself noted, his symptoms were often brought on by psychological stress, especially in social situations.

Darwin struggled with anxiety, depression and a profound fear of death throughout his adult life, and believed himself to be constitutionally prone to illness. On at least two occasions, depression incapacitated him for several months. He experienced numerous personal upsets during his life and found them difficult to cope with. The death of his father, the frequent and often serious illnesses of his children and his wife's numerous pregnancies caused him great anxiety and at times plunged him into prolonged bouts of depression. It is noteworthy that he also suffered from skin eruptions, which he referred to as eczema, since it is now recognized that chronic skin problems can be exacerbated by stress.

The eminent psychiatrist John Bowlby and other medical luminaries have proposed an attractive alternative to the simple Chagas' disease hypothesis, which interweaves both psychological and organic elements. They have suggested that Darwin's manifold symptoms are better explained by an anxiety disorder associated with abnormally rapid breathing, or hyperventilation.

Anxiety disorders can manifest themselves as panic attacks. The affected person feels an overwhelming but inexplicable sense of doom; the pulse races, the heart pounds, they tremble and feel short of breath. These symptoms are occasionally mistaken for heart disease – an assumption that Darwin and his doctors appear to have made. Hyperventilation is often associated with chronic anxiety and can be triggered by psychological stress. If it persists, hyperventilation can produce a wide range of physiological and psychological symptoms, including gastric disorders, flatulence, palpitations, nausea, blurred vision, fatigue, anxiety, dizziness, headaches and chest pains. It has been estimated that hyperventilation is a contributory factor in 10 per cent of all hospital outpatient referrals. The symptoms of hyperventilation arise from fluctuations in the concentration of carbon dioxide in the blood. Mild psychological stressors can tip a person who is prone to the condition into a state where their hyperventilation is sufficient to produce dramatic symptoms. All of this is consistent with Darwin's condition.

For many years Darwin had to struggle not only with his mysterious, debilitating infection, but also with the implication that he must be a neurotic hypochondriac or malingerer. As he wrote to his friend Sir Joseph Hooker in 1845:

> You are very kind in your inquiries about my health; I have nothing to say about it, being always much the same, some days better and some worse. I believe I have not had one whole day, or rather night, without my stomach having been greatly disordered, during the last three years, and most days great prostration of strength: thank you for your kindness; many of my friends, I believe, think me a hypochondriac.

The implication of hypochondria must have been distressing for a man whose life had been dedicated to scientific understanding. Like contemporary sufferers from chronic fatigue syndrome, Darwin was under great pressure to convince himself and others that his illness was genuine. Perhaps he did this by unconsciously exaggerating the severity of his symptoms. He may well have been encouraged in this by his devoted wife Emma, who cosseted him to the extent that some biographers have accused her of encouraging hypochondria.

Peter Medawar characteristically hit the nail on the head when he wrote this of Darwin's illness:

> The diagnoses of organic illness and of neurosis are not, of course, incompatible. Human beings cannot be straightforwardly ill like cats and mice; almost all chronic illness is surrounded by a penumbra of gloomy imaginings and by worries and fears that may have physical manifestations.

The real point is that Darwin's illness was neither purely psychological nor purely physical. There are grounds for believing that he might have had Chagas' disease, or some other undiagnosed organic disorder. But there are equally good grounds for believing that Darwin's psychological state had an important bearing on the course and expression of his disease, and that his illness was further complicated by his psychological reaction to it.

Parting shots

I hope I have convinced you of some simple but far-reaching truths. That our mental state and physical health are inexorably intertwined. That stress, depression and other psychological factors can alter our vulnerability to many diseases, including bacterial and viral infections, heart disease and cancer. That the relationship between mind and health is mediated both by our behaviour and by biological connections between the brain and the immune system. That these connections work in both directions, so our physical health can influence our mental state. That all illnesses have psychological and emotional consequences as well as causes. That there is nothing shameful or weak about the intrusion of thoughts and emotions into illness. That our social relationships with other people are central to health. And that our dualist habit of contrasting mind and body, as though they were two fundamentally different entities, is deeply misleading.

I hope too that you no longer feel (if indeed you ever did) that accepting the reality of mental influences on health means suspending your disbelief and signing up to some dubious brand of alternative medicine. There is nothing alternative or dubious about the scientific research I have described.

Naturally, many questions about the relationships between brain, behaviour, immunity and health are as yet unanswered. But we should not be surprised by this. After all, the human organism is the most complex and extraordinary entity in the known universe.

NOTES

1 The Body of Knowledge

1 It is estimated that up to thirteen Israelis died from suffocation through misusing their gas masks. Breathing difficulties may have been a factor in some of the unexpected deaths.

2 We have no cause to be smug or complacent about the decline in serious infectious diseases in industrialized nations. Their demise may be only temporary. Formerly curable infections are now staging a comeback, thanks to the rapid increase in strains of bacteria, viruses and parasites that are resistant to most (or all) standard drugs and antibiotics. Many hospitals, for example, are now plagued by strains of *Staphylococcus aureus* that are resistant to virtually all known antibiotics.

3 Lucetta Le Sueur in Thomas Hardy's *The Mayor of Casterbridge* is another fictional victim of shame. When her former liaison with the now disgraced Michael Henchard is revealed she is publicly humiliated by the burghers of Casterbridge. The highly strung Lucetta has an hysterical fit and dies. In Hardy's *The Woodlanders*, John South is bedridden and gripped with an obsession that he will be killed by an elm tree. He is. The offending elm tree is chopped down on his doctor's advice; when South notices that the tree is gone he has a fit and dies before the night is out.

4 Armand pays a heavy price. The distress he suffers after Marguerite's death is such that he insists on having her body exhumed so he can gaze upon her one last time. It is a shocking experience and the narrator warns Armand to take care, lest he kill himself with such emotions. It was sound advice: the emotional shock of seeing Marguerite's decomposing body plays havoc with Armand's health and he immediately falls seriously ill with 'brain fever'.

5 In Charlotte Brontë's *Jane Eyre*, the long-suffering Helen Burns succumbs to consumption amid the grim conditions of Lowood Asylum. The stoical, uncomplaining Helen welcomes her own death and her morbid wish is soon granted. The character is said to be an almost exact representation of Charlotte Brontë's sister Maria, who died of tuberculosis under similar circumstances at a religious boarding school in Yorkshire. Other victims of tubercular tragedies from this period include Mimi in Puccini's *La Bohème*, Little Eva in *Uncle Tom's Cabin* and Paul Dombey in *Dombey and Son*.

2 Shadows on the Sun

1 Claims that Haitian voodoo rites rely on nerve poisons rather than psychological manipulation are most likely unfounded (see Booth W. 'Voodoo science', *Science*, **240**, 274 (1988)).

2 Calculating a single composite score in this simple way assumes that lots of minor life events have the same overall impact as one really nasty life event. Not surprisingly, it isn't that simple.

3 To be precise, 394 of the subjects were given nasal drops containing one of five respiratory viruses; the remaining 26 subjects acted as controls and received inert saline nose drops.

4 A distinction is sometimes drawn between the signs and symptoms of a disease. Signs are things that are apparent to the physician but not the patient, such as the results of a blood test. Symptoms are what the patient experiences, such as pains. I shall use the words in their looser, colloquial sense.

Psyche's Machine: The Inside Story

1 More verbal niceties. The terms illness and sickness can have slightly different connotations. Some writers have suggested that 'illness' should be restricted to describing how the patient feels, while 'sickness' should refer only to the patient's overt behaviour, such as seeking medical assistance. According to this usage people can be ill but not sick, and vice versa. For the sake of simplicity I use the words more or less interchangeably in their looser, colloquial sense.

2 *The Sorrows of Young Werther* was a semi-autobiographical work, based partly on Goethe's own unrequited love for one Charlotte Buff. Goethe's much-imitated (and much-parodied) novel was an immediate bestseller throughout Europe. It inspired a wave of spin-offs: there were Wertheresque operas, poems, plays, novels, waxworks, songs, porcelain figures, jewellery and paintings. There was even an *Eau de Werther* perfume.

3 A vaccine against HIV must fulfil several demanding criteria if it is to be really effective and practicable on a worldwide scale. In addition to conferring life-long immunity against the virus in one shot, the vaccine must be cheap, easily administered (ideally by mouth), safe, stable when stored in adverse climates and effective against all strains of HIV. This is a very tall order.

4 *Jude the Obscure* contains other illuminating episodes, including a nice example of an emotionally-triggered illness. Even though she really loves Jude, Sue marries an older man – the schoolmaster Phillotson. However, Sue refuses to sleep with Phillotson and the unfortunate schoolmaster eventually agrees to let her go and live with Jude. A great scandal ensues and a riotous public meeting is held at which Phillotson's decision is roundly condemned:

> When Phillotson saw the blood running down the rector's face he deplored almost in groans the untoward and degrading circumstances, regretted that he had not resigned when called upon, and went home so ill that next morning he could not leave his bed. The farcical yet melancholy event was the beginning of a serious illness for him; and he lay in his lonely bed in the pathetic state of mind of a middle-aged man who perceives at length that his life, intellectual and domestic, is tending to failure and gloom.

Later in the story, Jude and Sue experience an enormous tragedy. 'Little Father Time', Jude's melancholic son from his brief first marriage, kills Jude's and Sue's other two children and hangs himself. Sue is heavily pregnant at the time. The shock of this horrific tragedy sends her into premature labour and her baby is still-born. Another of Hardy's creations who hastens his own death through self-neglect is Giles Winterborne in *The Woodlanders*. Giles is lying ill in his cottage in the woods when his former sweetheart, Grace, seeks shelter there. She is now another man's wife and Giles' stoicism and extreme propriety make him insist that Grace stay in his cottage while he sleeps outside under a woefully inadequate shelter. A storm rages and he grows weak with fever, yet still he declines Grace's invitation to join her in the hut. Giles becomes delirious and dies, a victim of his passive, resigned personality and rigid moral code.

5 *B*-lymphocytes (or B-cells) are so called because they are produced and develop in the *b*one marrow. (In birds, B-lymphocytes are produced in a different organ which, by a stroke of etymological luck, is called the *b*ursa.) *T*-lymphocytes mature in the *t*hymus and are stored mainly in the spleen and lymph nodes. About 80 per cent of circulating lymphocytes are T-lymphocytes and 5–15 per cent are B-lymphocytes. The remainder are non-T/non-B lymphocytes such as the natural killer cells. T-lymphocytes are further sub-divided into two main classes: CD4 cells (otherwise known as T4 or helper cells), which include helper-inducer T-cells and helper-suppressor T-cells; and CD8 cells (sometimes called T8 or suppressor/cytotoxic cells), which include cytotoxic T-cells and suppressor T-cells. Confused?

6 Immunoglobulins (Ig) are divided into five main classes, unhelpfully known as IgG, IgA, IgM, IgD and IgE. They have different functions. IgG, the principal immunoglobulin in blood serum, is a crucial element in the body's defences against bacterial infection. IgA is the principal immunoglobulin in mucous secretions such as saliva, tears and snot. IgM helps to fight infection by destroying bacteria or sticking them together. IgE is involved in allergic reactions. The role of IgD is rather obscure, but it is thought to have a receptor function on the surface of lymphocytes.

7 The mitogens most commonly used for assessing lymphocyte responsiveness are phytohaemagglutinin (PHA), concanavalin A (ConA) and pokeweed mitogen (PWM). PHA and ConA stimulate T-lymphocytes. PWM stimulates both T- and B-lymphocytes.

8 Much of the research in psychoneuroimmunology has so far focused on lymphocytes and NK cells. However, it has become evident that other components of the immune system, such as monocyte/macrophages, also play a major role in the mind–immunity connection.

9 For an introduction to the business of measuring behaviour, see P. Martin and P. Bateson *Measuring Behaviour*, 2nd edn, Cambridge University Press, 1993.

4 Mind and Immunity

1 A statistical method known as meta-analysis is used for drawing together all the available evidence on a particular phenomenon, to give the firmest possible conclusions. Meta-analysis involves collating and analysing the data from all the published scientific studies which meet certain criteria of objectivity and experimental design. Tracy Bennett Herbert and Sheldon Cohen (1993: see References) carried out a meta-analysis of the relationships between psychological stress and immunity in humans using data from thirty-eight published studies. Their meta-analysis confirmed that there are clear, consistent connections between various forms of psychological stress and diverse measures of immune function.

2 If a depressed person is later diagnosed as having a serious illness it is possible that the depression might have been an early symptom of the same illness. However, in this study the very long periods that elapsed between detecting depression and the subsequent development of cancer make it unlikely that depression was an early manifestation of undetected cancer.

3 One reason why the results of research into depression and immunity are not always consistent is that depression can mean different things to different scientists. Depression can range from a relatively mild drop in mood to a serious, debilitating mental illness requiring hospital treatment. Psychiatrists do not always agree in their diagnosis of depression. Moreover, different types of depression probably have different biological origins.

4 Dostoyevsky was the son of a physician. Like his creation Raskolnikov, Dostoyevsky was battling with the effects of poverty at the time he wrote *Crime and Punishment*. The book contains several instances of stress-related disorders; for example, at the start of Raskolnikov's trial his mother is stricken with fever and delirium, '[her] illness was of some strange, nervous kind and was accompanied by something close to insanity'. She dies two weeks later. Sonya's consumptive and impoverished mother goes mad in the street after suffering unbearable torments, including the death of her alcoholic husband and her eviction, along with her small children, from their rented room. Her tubercular condition is exacerbated by the appalling stress and she dies. Towards the end of the story, when Raskolnikov is in prison, he becomes estranged from all the other prisoners and is taken ill – seemingly because of an emotional collapse: 'He had been ill for a long time; but it was not the horrors of life in penal servitude . . . that had worn him down . . . it was wounded pride that had made him fall ill.'

5 Pavlovian conditioning is a form of associative learning. The other main type of conditioning is operant (or instrumental) conditioning, in which the animal learns that a particular behaviour pattern (such as pressing a lever) produces a rewarding outcome (such as food) or an aversive outcome (such as electric shock). Conditioned responses are reversible. If the experimenter repeatedly rings the bell without giving the dog any food, the dog soon learns not to salivate. This decline in the conditioned response is referred to as extinction. Immune conditioning is extinguished if the biologically neutral stimulus is repeatedly presented without the immunologically active stimulus.

6 Learned taste aversion is one of the many tricks nature uses to keep animals alive in a dangerous world. If an animal tries a novel-tasting food and then gets sick it will develop a long-lasting aversion to that food and will avoid eating it thereafter. What is especially interesting about learned taste aversion is that the animal can readily learn to associate two stimuli (the novel taste and the subsequent nausea) even though they are separated in time by several hours. In the real world, effect normally follows cause with little or no delay; the sound of a predator means that a predator may appear imminently rather than tomorrow. Conditioning is therefore geared to learning about associations between events that occur close together in time. If there is a lengthy delay between pressing a lever and getting food, a rat cannot learn to associate the two. But some cause–effect relationships in nature *do*

have long delays built into them; if a rat eats something poisonous, hours rather than seconds will normally elapse before it starts to feel the effects. Learned taste aversion is specific to tastes and illness; you cannot, for example, teach a rat to avoid a visual stimulus or a sound that is followed hours later by sickness.

7 Rest assured that Ader and Cohen's experiment was impeccably designed, with numerous control groups. Subsequent experiments on immune conditioning, of which there have been many, have used even more elaborate experimental designs to ensure that every conceivable confounding factor has been taken into account.

8 The immune reaction to tuberculin is an example of the delayed-type hypersensitivity response.

9 One reason for the inconsistency in the data is that handedness has been defined in different ways by different researchers. Handedness varies along a continuous spectrum from complete left-handedness to complete right-handedness. It makes a considerable difference whether you are talking about a person who is 100 per cent left-handed, as opposed to one who does not always use their right hand.

5 The Demon Stress

1 Just to confuse you, some scientists used to refer to the outside events (i.e., the stressors) as 'stress' and the stress response as 'distress'.

2 People suffering from Addison's disease have abnormally low levels of cortisol. One of the lesser known consequences of Addison's disease is its remarkable effect on sufferers' senses. Their ability to detect weak sensory stimuli may be increased by a factor of ten thousand or more above normal, while their ability to recognize and discriminate between different stimuli is severely impaired. Conversely, people suffering from Cushing's syndrome, who have abnormally high cortisol levels, experience the opposite: their sensory acuity is impaired, but their discrimination is improved. The relationship between sensory perception and cortisol level also shows up in the daily fluctuations in hormone levels that occur naturally in all of us. Irrespective of stress, our blood cortisol level varies by a factor of two or three over the twenty-four-hour period, according to a regular circadian rhythm. Cortisol peaks in the small hours of the morning when we are asleep and is at its lowest in the late afternoon and early evening. Our sensory abilities also vary rhythmically, in parallel with these changes in cortisol.

3 Humans have two adrenal glands. They are roughly triangular in shape and sit immediately above the kidneys (hence their other name, the suprarenal glands). Each adrenal gland has two distinct parts. The outer shell is known as the adrenal cortex and secretes various steroid hormones. The inner core is known as the adrenal medulla. Adrenaline (otherwise known as epinephrine) is secreted by the adrenal medulla. Noradrenaline (alias norepinephrine) is secreted from sympathetic nervous system nerve endings and the adrenal medulla. Both adrenaline and noradrenaline are examples of a class of chemical messengers known as catecholamines.

4 To give two examples of the numerous control loops: adrenaline and noradrenaline stimulate the release of ACTH from the pituitary, while glucocorticoids released from the adrenal cortex regulate the production of adrenaline and noradrenaline, and inhibit ACTH secretion. It's all very complicated and extremely clever.

5 The various steroid hormones released by the adrenal cortex are

known collectively as corticosteroids. The two main categories of corticosteroid hormones are the glucocorticoids, of which cortisol (or hydrocortisone) is the prime example in humans, and the mineralocorticoids, of which aldosterone is an example.

6 Like many other biological functions, the immune system exhibits a daily (circadian) rhythm in its activity. Diverse aspects of immune function fluctuate regularly over the course of the twenty-four-hour sleep-wake cycle. Cortisol levels also vary rhythmically, but in opposite phase to immune function. The latter is highest when cortisol is lowest and vice versa. Our susceptibility to bacterial infection also fluctuates over the twenty-four-hour period, in synchrony with the variations in immune function and cortisol level.

7 The hopelessness and helplessness of a man who believes himself incapable of altering his fate is starkly portrayed in Ernest Hemingway's story 'The Killers'. Two hit men burst into a small-town diner looking for Ole Andreson, a former heavyweight prizefighter. They intend to kill Andreson when he comes in for his supper. Luckily, Andreson doesn't turn up that night and the killers leave. Someone from the diner runs to warn Andreson of the danger, but he merely lies on his bed, unable to motivate himself to do anything about it. He rejects the suggestion that the police should be informed. Though he could probably evade his would-be killers, Andreson opts to lie back and await his death. As far as he is concerned the killers have caught up with him and death is the inevitable outcome.

8 Of course, this strategy only works within certain limits. There is no suggestion that severe or prolonged stress may be beneficial. People who have experienced extreme stress – such as concentration camp survivors – are often found to have suffered lasting psychological damage. One consequence may be that they are less able to cope with other stressors later in life. For example, a study of Israeli Holocaust survivors during the Gulf War showed that their reaction to a past severe stressor (the Holocaust) could be reactivated decades later by a new stressor (Iraqi missile attacks on Israel). Holocaust survivors whose homes were damaged by Iraqi missiles suffered unusually high levels of psychological trauma.

6 Other People

1 The cruelty of a social group towards certain of its members is also a theme of William Golding's first novel, *Lord of the Flies.* This tells the story of a group of schoolboys marooned on an uninhabited tropical island after a plane crash. The veneer of civilization is soon peeled away, the social structure disintegrates and the boys regress into savagery. They mercilessly turn on certain individuals. One boy is torn apart. Another is beaten. The fat, unattractive Piggy is stoned, then killed (and probably eaten).

2 I should add that there does not appear to be anything intrinsically magical about the marriage ceremony itself; what really matters for health is being part of a close, long-term, loving relationship. For a growing number of people this does not necessarily entail a formal marriage.

3 In this particular study, socially isolated men were not found to be at significantly greater risk of developing cancer. However, they did have a poorer chance of surviving cancer if they developed it.

4 Social scientists use the term social network to refer to the entire group

with whom an individual has some form of regular social contact. The social network includes friends, relatives and colleagues as well as immediate family. The help these people provide (or are perceived to provide) to the individual, in whatever form, is referred to as social support.

5 The cross-species benefits of social interactions can work in the opposite direction as well. Experimental studies with animals have shown that the reassuring presence of a familiar human can increase an animal's resistance to disease. For instance, one study found that interacting with humans counteracted the development of heart disease in rabbits. Two groups of rabbits were fed a cholesterol-rich diet which produces atherosclerosis. Rabbits in one group were visited several times a day, talked to, stroked and played with by a familiar person, while a control group received no extra human attention. Six weeks later the rabbits were killed, autopsied and examined for signs of heart disease. Remarkably, the rabbits who had interacted daily with a human were found to have 60 per cent less damage to their coronary arteries.

7 The Wages of Work

1 The statistics in the first part of this chapter are culled from a variety of sources, including the UK Health And Safety Executive, the UK Social Attitudes Survey, Harris Research, Demos and the London Hazards Centre. Unless otherwise stated the figures relate to the early or mid 1990s.

2 To complicate matters slightly, civil servants in the lowest grades also had higher rates of smoking, obesity, high blood pressure and physical inactivity than those in senior grades. However, even when these other risk factors were removed statistically the data still showed a strong connection between job grade and health.

3 Statistics indicate that working women also have lower mortality rates and better health than housewives, some of whom have presumably chosen not to work.

8 Sick at Heart

1 Having a lot of cholesterol in your food and a lot of cholesterol in your blood are two different things that should not be confused. The evidence that variations in dietary cholesterol intake have much impact on blood cholesterol levels (and hence heart disease) is by no means conclusive. A very high level of blood cholesterol – which *is* a risk factor for heart disease – often results from a genetic defect or other disorder, and not from eating fatty food.

2 The dividing line between high and normal blood pressure is essentially arbitrary and definitions of what constitutes hypertension vary. Normal blood pressure also changes with age. However, for someone in an industrialized nation like Britain, normal blood pressure is in the range 100–140 mmHg (systolic) over 60–90 mmHg (diastolic), with an average of 120/80. According to the World Health Organization, anyone whose blood pressure is consistently more than 160/95 has hypertension.

3 In another experiment, two patients who were exposed to mild psychological stressors in the laboratory suffered acute pulmonary oedema (a sudden build-up of fluid in the lung which often results from heart failure).

4 As we saw in chapter 7, a classic study of British civil servants found that those in the lowest job grades were more than three times as likely to die from coronary heart disease as those in

the highest grades. The high-risk junior staff also had the highest levels of fibrinogen in their blood. It has been suggested that stress-related differences in fibrinogen levels might contribute to the general grade-related differences in heart disease.

5 This is not the appropriate place for a treatise on the distinction between personality and behaviour. I am using the terms in their colloquial rather than technical senses. Behaviour refers to what people *do*, their overt actions; it can be observed and objectively measured by others. Personality refers to the sum total of psychological, emotional and behavioural characteristics which distinguish one individual from another. Thus, personality includes distinctive styles of thought and emotional response as well as actions. Theories about the mind and heart disease are primarily concerned with personality in the broad sense rather than just overt behaviour patterns, since the way in which individuals think and respond emotionally is important, too. But for practical purposes it is usually the overt behaviour patterns that scientists measure. Hence there is a tendency (reflected in my account) to use the terms personality and behaviour more or less interchangeably.

6 To complicate matters, the Type A men also exhibited more of the conventional risk factors for heart disease. They had much higher blood cholesterol levels and were more likely to smoke, have a family history of heart disease, work longer hours and sleep less. We shall return to this point later.

7 Another character who displays the explosive, exaggerated speech patterns of a Type A is Pistol, King Henry's former drinking crony in Shakespeare's *Henry V*. The name says much about his character: 'For Pistol, he hath a killing tongue and a quiet sword.' This is a man of words rather than actions. And his words are mostly boastful, blustering and bombastic. His speech is full of alliteration and gallumphing rhythms, peppered with pompous phrases, with the vainglorious Pistol all the while 'swelling like a turkey-cock'. Pistol is easily roused to anger. When he argues with his old friend Nym his insults are among the choicest ever uttered in the English language; 'Pish for thee, Iceland dog, thou prick-eared cur of Iceland' constitutes a mild rebuke by Pistol's standards.

8 The interviewer looks for a range of standard indicators of Type A personality. These reflect both the content of the subject's answers and the subject's behaviour during the interview. The interviewer is on the look-out for characteristics such as loud, emphatic, hurried speech; hostility; impatience; abrupt gestures; a sharp response to mild provocations or irritations; and a desire to exert control over the social situation.

9 The various paper-and-pencil techniques, such as the Jenkins Activity Survey, the Framingham Type A Scale and the Bortner Rating Scale, appear to measure different things and do not agree with one another in all respects. Some methods are better at picking out impatience or competitiveness, while others are better at detecting hostility.

10 Other factors may have contributed to inconsistencies in the research data. For a start, the relationship between Type A behaviour and heart disease probably varies according to which social group is studied. Most of the original research was conducted on white, middle-class, middle-aged American men; the relationship may be different for women, children or other ethnic and socioeconomic groups. Another complication is that behaviour changes over time, so a person who was assessed as Type A

might no longer be Type A ten or twenty years later. This could explain the drop-off in correlation between Type A and heart disease in certain long-term studies. In addition, the impact of Type A behaviour can change over time as the individual's circumstances change; for example, Type A behaviour might have a greater impact in the early or middle years of life, when people are under greater stress at work and at home, than in later years. A further complication that might be relevant to our story is the general decline in heart disease that has been taking place in industrialized societies over the past twenty or thirty years. Even in Britain, which has rather lagged behind in this respect, the death rate from heart disease has started to fall (among some sections of the population at least). As people have become more aware of the risks, some of them have responded by not smoking, eating less saturated fat or taking more exercise. It is conceivable that these long-term changes in behaviour and the associated decline in heart disease may have blurred the connection with Type A behaviour.

9 *The Mind of the Crab*

1 No one has actually counted them all, of course, and estimates do differ. The true figure might be nearer 10^{14}, but what's a factor of ten between friends?

2 To be strictly accurate, not *all* of our cells have a full set of genes: red blood cells have no nucleus.

3 Malignant melanoma is a tumour of the melanocytes: the cells in the epidermis, or outer layer of skin, which produce the dark pigment melanin. Malignant melanoma is particularly virulent because of its propensity to metastasize – that is, to spread secondary tumours to other parts of the body.

4 Most of the evidence for links between psychological factors and cancer in humans is correlational; that is, scientists observe a statistical association between a psychological factor (such as stress or depression) and the subsequent incidence or progression of cancer. Ethical considerations obviously rule out the possibility of experimentally inducing tumours in human subjects, or subjecting them to severe stress in the expectation that they will get cancer. In fact, ethical considerations make such experiments increasingly questionable when conducted on other species as well.

5 A stressor's duration also affects its impact on tumours. In one experiment, rats which had been given a carcinogenic chemical and then intermittently stressed over a three-month period developed more tumours. However, if the same stressor regime was applied for five months it reduced the number and size of tumours.

6 Before she was overshadowed by Virginia Woolf, May Sinclair was one of the most highly regarded women novelists in Britain. Her work was even compared with Charlotte Brontë's. *The Life and Death of Harriett Frean* appeared in 1922, at a time when Freud's psychoanalytic theories were influential among British intellectuals. Sinclair drew heavily on the Freudian notions of repression and sublimation in *Harriett Frean* and her other books. As well as the Type C Harriett and her cancer, Sinclair's novel touches on a variety of relevant themes. For instance, Harriett demonstrates the distinction between sickness-related behaviour and true disease. During a bout of pleurisy several years before her

death, she discovers the pleasures of the sickness role:

> Out of the peace of illness she entered on the misery and long labour of convalescence . . . She didn't want to get well. She could see nothing in recovery but the end of privilege and prestige . . .

Elsewhere in the story, Harriett's best friend Priscilla develops hysterical paralysis, a classic psychosomatic illness in the old and dubious sense of the word. Doctors can find no organic cause for the paralysis: the only reason why Priscilla can't walk is that she can't walk. It is implied that the true origins of the disorder lie in Priscilla's relationship with her unloving husband Robin:

> Robin wasn't in love with her, and she knew it. She developed that illness so that she might have a hold on him, get his attention fastened on her somehow. I don't say she could help it. She couldn't.

Later on, Harriett's father suddenly loses all his money on the stock exchange. In traditional literary style, the shock sends him into severe decline and he dies of heart disease.

7 Arnold Bennett himself had mixed feelings about the impact of his emotions on his physical state. Bennett suffered from insomnia and, like his creation Earlforward, regarded himself as 'highly strung'.

8 Almost inevitably, given the complexity of the underlying issues, the scientific literature also contains examples of studies that failed to find any significant connection between psychological factors and survival from cancer. One of the most widely quoted of these negative studies has been criticized because it was based on a sample of patients who were in the terminal stages of cancer. When someone is already on the verge of death it is inherently less likely that psychological factors – or anything else for that matter – will make much difference to them.

10 *Encumbered with Remedies*

1 A conditioned association with the distinctive perfume habitually worn by a sexual partner produces anything but a physiological relaxation response.

2 In the Harvard study vigorous exercise was defined as activity in which metabolic rate rose to at least six times resting level – in other words, pretty vigorous. The Harvard researchers reassuringly pointed out that although non-vigorous exercise was not significantly correlated with longevity in their study it has been shown to have other health benefits. The study of Israeli men, however, did find clear benefits accruing from physical activity that was neither intense nor particularly frequent.

3 A retrovirus (of which HIV is just one example) is a virus whose genes are encoded in the form of RNA, rather than the DNA which is the universal genetic material in animals and plants. Retroviruses use an enzyme called reverse transcriptase to transcribe their genes into the standard DNA format before incorporating them into the host's genetic material.

4 In 1954 the Catholic Church set up the International Medical Committee of Lourdes (CMIL), an international body of Catholic doctors whose ultimate responsibility it is to judge the authenticity of supposed miracle cures. Over the following thirty years the CMIL concluded that only nineteen cases were medically inexplicable, of which thirteen were officially recognized by the Vatican as miracles.

5 The men's blood pressure was compared against that of a control group of hypertensives who were not informed about their condition. Subjects were assigned randomly to the two groups.

11 Exorcising the Ghost in the Machine

1 There are numerous variations on the basic themes of dualism and monism. Both are broad philosophical churches in their own right, encompassing many shades and complexities of opinion. Hard-line dualists such as Wittgenstein and Leibniz argued that the mind and body are completely separate and independent of one another. Less extreme dualists, whilst maintaining the fundamental distinction between mind and body, accepted that the two interact in important ways. Plato, St Augustine and St Thomas Aquinas all believed that the mind is the body's animating and controlling spirit but is nonetheless made of different stuff.

2 Hard-line monists such as Berkeley, Hegel and Teilhard de Chardin have argued that all material experience springs from the mind, in effect subsuming the physical body within the mind. Bishop Berkeley famously maintained that material objects only exist when they are perceived by a mind, to which Samuel Johnson famously riposted 'I refute it *thus*' and kicked a large stone. At the other end of the monist spectrum are those taking a strongly materialistic stance, in effect subsuming the mind within the body. Epicurus, Hobbes and Diderot regarded the mind as an essentially physical phenomenon. The seventeenth-century Dutch philosopher Spinoza was a particularly trenchant critic of dualism. The strongly materialistic and reductionist form of monism reached its zenith in the mid twentieth century with the behaviourist school of experimental psychology, founded by J. B. Watson and B. F. Skinner. By focusing solely on observable patterns of behaviour, the behaviourists in effect denied the very existence of the mind.

3 The monist views of Hippocrates and Epicurus were not shared by all the great thinkers of ancient Greece. Plato took a predominantly dualist stance in his earlier works, arguing that the body and soul are inter-dependent but fundamentally distinct. Plato maintained that inside each of us is an immortal and immaterial soul which gives life to the body, but which is nonetheless quite distinct from the body. In his later dialogues, however, Plato displayed more monist tendencies, describing the soul, or mind, as an attribute of the body rather than something entirely separate from it. Plato's pupil Aristotle was even less of a clear-cut dualist than his mentor. In works such as *De Anima*, Aristotle argued that the *psyche* is the 'form' or organizing essence of a living organism – the thing that imbues it with life and purpose. As far as Aristotle was concerned, the *psyche* could not be separated from the physical body (or *soma*) any more than the property of sight could be separated from the eye. Yet in other respects Aristotle remained a dualist. He argued that conscious thought is wholly separate from the physical body. To Aristotle, like Plato, the *psyche* might reside in a physical body, but is not itself a physical entity.

4 Lest I be accused of Eurocentrism, let me add that science and mathematics were kept alive in Arab cultures.

5 The thirteenth-century theologian and philosopher Thomas Aquinas (1225–74) devoted much of his life to reconciling Aristotle's scientific writings with Christian beliefs. Aquinas

concluded that the human soul is entirely separate from the physical body, that it comes from God, and that it is immortal.

6 Queen Kristina (1626–89) is a fascinating figure in her own right. Highly intelligent and reportedly very beautiful, she published several works on philosophy and ethics and was an active patron of the arts and scholarship. She was also bisexual and refused to marry. Kristina abdicated from the throne of Sweden in 1654, converted to Catholicism and ultimately became an atheist.

7 Descartes was not the only scholar for whom a visit to Queen Kristina's court proved fatal. Hugo Grotius, the Dutch philosopher, lawyer and theologian, visited Kristina and, like Descartes, succumbed to a respiratory infection. He died on his way back to Holland in 1645.

12 A Fresh Pair of Lenses

1 'How does it work?' (the analysis of mechanism) is sometimes referred to as the question of proximate causation. 'How did it evolve?' (the analysis of evolutionary origins and adaptive function) is rather grandiosely referred to as the question of ultimate causation. More prosaically these two varieties of question are referred to as 'how' and 'why' questions.

2 Another biological term for development is ontogeny. To complicate matters further, the developmental processes that are responsible for the assembly of an adult organism are themselves the products of natural selection. They evolved to make individuals better able to survive and reproduce in a harsh world. The ability of animals to modify their behaviour by learning is one example of a developmental process that has been shaped by natural selection.

3 In the case of humans it would be more accurate to refer to the caregiver–offspring (rather than mother–offspring) relationship.

4 The experiment employed a double-blind, randomized, placebo-controlled design. Double-blind means that neither the subjects nor the researchers knew which drug (if any) they had taken until after the trial was over, thereby avoiding unconscious bias. Randomized means that subjects were randomly assigned to the various groups, again to avoid bias. And placebo-controlled means that the effects of taking a drug were compared against a placebo.

5 Morning sickness is something of a misnomer because it does not only occur in the mornings. Pregnant women can experience taste aversions, nausea and vomiting at any time of the day. Moreover, many women develop taste aversions without necessarily experiencing acute nausea or vomiting.

6 Profet has applied a similar evolutionary analysis to the increasingly problematic phenomenon of allergy, which now affects as many as one in four people in industrialized societies. She has constructed a compelling theory that allergy is a last-ditch immunological defence against toxic substances, and is designed to expel toxins from the body as rapidly as possible when all else has failed. The idea that allergies might have a protective function is consistent with tentative evidence that those who suffer from allergies have a slightly reduced risk of developing certain forms of cancer. The evolutionary perspective suggests yet again that we should pause for thought before indiscriminately using drugs to suppress allergies; the blithe assumption that allergies have no benefits, only costs, may turn out to be mistaken.

7 If anything, this is an over-estimate

of the number of generations, since it assumes an average reproductive cycle (from birth to reproduction) of 20 years, making a hundred human generations during the 2,000 years since the birth of Jesus. Eighty generations of 25 years each may be nearer the mark.

8 In his private autobiography Darwin described his religious views:

> Thus disbelief crept over me at a very slow rate, but was at last complete ... I can indeed hardly see how anyone ought to wish Christianity to be true; for if so, the plain language of the text seems to show that the men who do not believe, and this would include my Father, Brother and almost all my best friends, will be everlastingly punished. And this is a damnable doctrine.

REFERENCES

Even if you have no intention of pursuing any of these references you may still wish to skim through the list. It should at least convey some impression of the sheer scope of the published material. I could not begin to cite more than a fraction of the scientific papers and books that have been published on this subject. My aim here is to give a representative flavour, with a bias towards the most recent material and reviews.

To save trees I have cited references in a truncated form. Scientific papers often have several authors. Unless you are one of them you are probably not interested in who they all are, so I have named only the first author; the rest are subsumed under the anonymity of 'et al'. In many cases I have also taken the liberty of truncating the title of a paper. The abbreviation 'NK' is used throughout for natural killer (cell). Journal names also appear in an abbreviated form. Thus the *New England Journal of Medicine* becomes *N. Engl. J. Med.*, while *Social Science and Medicine* becomes *Soc. Sci. Med.* In keeping with convention, the relevant volume number of the journal is in bold type and is followed by the first page number of the paper. Thus, '*J. Behav. Med.*, **16**, 143' indicates that the paper in question starts on page 143 of volume 16 of *The Journal of Behavioral Medicine*.

1 *The Body of Knowledge*

Angell M. 'Disease as a reflection of the psyche', *N. Engl. J. Med.*, **312**, 1570 (1985)

Carmel S. et al 'Coping with the Gulf War', *Soc. Sci. Med.*, **37**, 1481 (1993)

Engel G. L. 'The need for a new medical model', *Science*, **196**, 129 (1977)

Hall J. G. 'Emotion and immunity', *Lancet*, **2**, 326 (1985)

Kark J. D. et al 'Iraqi missile attacks on Israel. The association of mortality with a life-threatening stressor', *J. Am. Med. Assoc.*, **273**, 1208 (1995)

Keynes W. M. 'Medical response to mental stress', *J. Roy. Soc. Med.*, **87**, 536 (1994)

Lerner B. H. 'Can stress cause disease? Revisiting tuberculosis', *Ann. Intern. Med.*, **124**, 673 (1996)

Moran M. G. 'Psychiatric aspects of tuberculosis', *Adv. Psychosom. Med.*, **14**, 109 (1985)

Phillips D. P. & Smith D. G. 'Postponement of death until symbolically meaningful occasions', *J. Am. Med. Assoc.*, **263**, 1947 (1990)

Ramchandani D. et al 'Evolving concepts of psychopathology in inflammatory bowel disease', *Med. Clin. North Am.*, **78**, 1321 (1994).

Sarafino E. P. *Health Psychology* 2nd edn (NY, Wiley, 1994)

Solomon G. F. 'Whither psychoneuroimmunology?' *Brain Behav. Immun.*, **7**, 352 (1993)

Weiss D. W. et al 'Psychological and immunological parameters in Israelis during Scud missile attacks', *Behav. Med.*, **22**, 5 (1996)

CHRONIC FATIGUE SYNDROME (CFS)

Bates D. W. et al 'Clinical laboratory test findings in patients with CFS', *Arch. Intern. Med.*, **155**, 97 (1995)

Edwards R. H. et al 'Muscle histopathology and physiology in CFS', *Ciba Found. Symp.*, **173**, 102 (1993)

Epstein K. R. 'The chronically fatigued patient', *Med. Clin. North Am.*, **79**, 315 (1995)

Fukuda K. et al 'CFS', *Ann. Intern. Med.*, **121**, 953 (1994)

Gunn W. J. et al 'Epidemiology of CFS: the Centers For Disease Control Study', *Ciba Found. Symp.*, **173**, 83 (1993)

Krupp L. B. & Pollina D. 'Neuroimmune and neuropsychiatric aspects of CFS', *Adv. Neuroimmunol.*, **6**, 155 (1996)

Levy J. A. 'Viral studies of CFS', *Clin. Infect. Dis.*, **18**, S-117 (1994)

Lloyd A. R. et al 'Immunity and the pathophysiology of CFS', *Ciba Found. Symp.*, **173**, 176 (1993)

Maclean G. & Wessely S. 'Professional and popular views of CFS', *Br. Med. J.*, **308**, 776 (1994)

Mawle A. C. et al 'Immune responses associated with CFS', *J. Infect. Dis.*, **175**, 136 (1997)

Natelson B. H. et al 'Brain magnetic resonance imaging in patients with CFS', *J. Neurol. Sci.*, **120**, 213 (1993)

Richman J. A. et al 'CFS', *Am. J. Public Health*, **84**, 282 (1994)

Rowe P. C. et al 'Is neurally mediated hypotension an unrecognised cause of chronic fatigue?' *Lancet*, **345**, 623 (1995)

Saisch S. G. et al 'Hyperventilation and CFS', *Q. J. Med.*, **87**, 63 (1994)

Schweitzer R. et al 'Illness behaviour of patients with CFS', *J. Psychosom. Res.*, **38**, 41 (1994)

Sharpe M. et al 'Cognitive behaviour therapy for CFS', *Br. Med. J.*, **312**, 22 (1996)

Shorter E. 'Chronic fatigue in historical perspective', *Ciba Found. Symp.*, **173**, 6 (1993)

2 *Shadows on the Sun*

General

Cohen S. & Herbert T. B. 'Health psychology: psychological factors and physical disease', *Ann. Rev. Psychol.*, **47**, 113 (1996)

Cohen S. & Williamson G. M. 'Stress and infectious disease in humans,' *Psychol. Bull.*, **109**, 5 (1991)

Sheridan J. F. et al 'Psychoneuroimmunology: stress effects on pathogenesis and immunity', *Clin. Microbiol. Rev.*, **7**, 200 (1994)

Totman R. *Mind, Stress and Health* (London, Souvenir Press, 1990)

Death, disaster and voodoo

Appels A. & Otten F. 'Exhaustion as precursor of cardiac death', *Br. J. Clin. Psychol.*, **31**, 351 (1992)

Binik Y. M. 'Psychosocial predictors of sudden death', *Soc. Sci. Med.*, **20**, 667 (1985)

Cannon W. 'Voodoo death', *Psychosom. Med.*, **19**, 183 (1957)

Carmelli D. et al 'Twenty-seven-year mortality in the Western Collaborative Group Study,' *J. Clin. Epidemiol.*, **44**, 1341 (1991)

Dobson A. J. et al 'Heart attacks and the Newcastle earthquake', *Med. J. Austr.*, **155**, 757 (1991)

Engel G. 'Sudden and rapid death during psychological stress', *Ann. Intern. Med.*, **74**, 771 (1971)

Kamarck T. & Jennings J. R. 'Biobehavioral factors in sudden cardiac death', *Psychol. Bull.*, **109**, 42 (1991)

Katsouyanni K. et al 'Earthquake-related stress and cardiac mortality', *Int. J. Epidemiol.*, **15**, 326 (1986)

Morse D. R. et al 'Psychosomatically induced death', *Stress Med.*, **7**, 213 (1991)

Myers A. & Dewar H. A. 'Circumstances attending 100 sudden deaths from coronary artery disease', *Br. Heart J.*, **37**, 1133 (1975)

Natelson B. H. & Chang Q. 'Sudden death. A neurocardiologic phenomenon', *Neurol. Clin.*, **11**, 293 (1993)

Siltanen P. 'Stress, coronary disease, and coronary death', *Ann. Clin. Res.*, **19**, 96 (1987)

Trichopoulos D. et al 'Psychological stress and fatal heart attack', *Lancet*, **1**, 441 (1983)

TROUBLE, STRIFE AND SICKNESS

Boyce W. T. et al 'Influence of life events and family routines on childhood respiratory tract illness', *Pediatrics*, **60**, 609 (1977)

Chao C. C. et al 'Effects of immobilization stress on acute murine toxoplasmosis', *Brain Behav. Immun.*, **4**, 162 (1990)

Eysenck H. J. 'Personality, stress and cancer', *Br. J. Med. Psychol.*, **61**, 57 (1988)

Gross W. B. 'Effect of social stress severity on *E. coli* infection', *Am. J. Vet. Res.*, **45**, 2074 (1984)

Hamilton D. R. 'Immunosuppressive effects of predator induced stress in mice', *J. Psychosom. Res.*, **18**, 143 (1974)

Jemmott J. B. & Locke S. E. 'Psychosocial factors and human susceptibility to infectious diseases', *Psychol. Bull.*, **95**, 78 (1984)

Jonas B. S. et al 'Are symptoms of anxiety and depression risk factors for hypertension?' *Arch. Fam. Med.*, **6**, 43 (1997)

Kasl S. V. et al 'Psychosocial risk factors in the development of infectious mononucleosis', *Psychosom. Med.*, **41**, 445 (1979)

Kiecolt-Glaser J. K. et al 'Slowing of wound healing by psychological stress', *Lancet*, **346**, 1194 (1995)

Meyer R. J. & Haggerty R. J. 'Streptococcal infections in families', *Pediatrics*, **29**, 539 (1962)

Rahe R. H. 'Anxiety and physical illness', *J. Clin. Psychiatry*, **49**, 26s (1988)

Rosengren A. et al 'Psychological stress and coronary artery disease', *Am. J. Cardiol.*, **68**, 1171 (1991)

Russek L. G. et al 'The Harvard Mastery Of Stress Study 35-year follow-up', *Psychosom. Med.*, **52**, 271 (1990)

Somervell P. D. et al 'Psychologic distress as a predictor of mortality', *Am. J. Epidemiol.*, **130**, 1013 (1989)

Vaillant G. E. 'Effects of mental health on physical health', *N. Engl. J. Med.*, **301**, 1249 (1979)

Vogt T. et al 'Mental health status as a predictor of morbidity and mortality', *Am. J. Public Health*, **84**, 227 (1994)

LIFE EVENTS

DeLongis A. et al 'Relationship of daily hassles, uplifts, and major life events to health status', *Health Psychol.*, **1**, 119 (1982)

Lobel M. et al 'Prenatal maternal stress and prematurity', *Health Psychol.*, **11**, 32 (1992)

Pagel M. D. et al 'Psychosocial influences on new born outcomes', *Soc. Sci. Med.*, **30**, 597 (1990)

Rabkin J. G. & Struening E. L. 'Life events, stress, and illness', *Science*, **194**, 1013 (1976)

Rahe R. H. & Arthur R. J. 'Life change and illness studies', *J. Hum. Stress*, **4**, 3 (1978)
Sarason I. G. et al 'Life events, social support, and illness', *Psychosom. Med.*, **47**, 156 (1985)

THE MIND AND THE COMMON COLD

Broadbent D. E. et al 'The prediction of experimental colds in volunteers by psychological factors', *J. Psychosom. Res.*, **28**, 511 (1984)
Canter A. 'Changes in mood during incubation of acute febrile disease and the effects of pre-exposure psychologic status', *Psychosom. Med.*, **34**, 424 (1972)
Cohen S. et al 'Psychological stress and susceptibility to the common cold', *N. Engl. J. Med.*, **325**, 606 (1991)
Cohen S. et al 'Negative life events, perceived stress, negative affect, and susceptibility to the common cold', *J. Pers. Soc. Psychol.*, **64**, 131 (1993)
Evans P. D. et al 'Minor infection, minor life events and the four day desirability dip', *J. Psychosom. Res.*, **32**, 533 (1988)
Graham N. M. H. et al 'Stress and acute respiratory infection', *Am. J. Epidemiol.*, **124**, 389 (1986)
Stone A. A. et al 'Cold symptoms following rhinovirus infection and prior stressful life events', *Behav. Med.*, **18**, 115 (1992)
Totman R. et al 'Predicting experimental colds in volunteers from different measures of life stress', *J. Psychosom. Res.*, **24**, 155 (1980)

3 *Psyche's Machine: The Inside Story*

GENERAL

Adler N. & Matthews K. 'Health psychology: why do some people get sick and some stay well?' *Ann. Rev. Psychol.*, **45**, 229 (1994)
Farrar W. L. et al 'The immune logical brain', *Immunol. Rev.*, **100**, 361 (1987)
Kiecolt-Glaser J. K. & Glaser R. 'Psychoneuroimmunology and health consequences', *Psychosom. Med.*, **57**, 269 (1995)
Kropiunigg U. 'Basics in psychoneuroimmunology', *Ann. Med.*, **25**, 473 (1993)
Pennisi E. 'Neuroimmunology. Tracing molecules that make the brain-body connection', *Science*, **275**, 930 (1997)
Reichlin S. 'Neuroendocrine-immune interactions', *N. Engl. J. Med.*, **329**, 1246 (1993)
Sternberg E. M. et al 'The stress response and the regulation of inflammatory disease', *Ann. Intern. Med.*, **117**, 854 (1992)

THE PERCEPTION OF SICKNESS

Green S. A. *Mind and Body: the Psychology of Physical Illness* (Washington DC, American Psychiatric Press, 1985)

Kaplan G. A. & Comacho T. 'Perceived health and mortality', *Am. J. Epidemiol.*, 117, 292 (1983)

Mayou R. & Sharpe M. 'Diagnosis, disease and illness', *Q. J. Med.*, **88**, 827 (1995)

Mechanic D. 'Social psychologic factors affecting the presentation of bodily complaints', *N. Engl. J. Med.*, **286**, 1132 (1972)

Sharpe M. et al 'Why do doctors find some patients difficult to help?' *Q. J. Med.*, **87**, 187 (1994)

Shorter E. *From Paralysis to Fatigue. A History of Psychosomatic Illness in the Modern Era* (NY, Free Press, 1992)

BAD BEHAVIOUR

Aggleton P. et al 'Risk everything? Risk behavior, behavior change, and AIDS', *Science*, **265**, 341 (1994)

Anda R. F. et al 'Depression and the dynamics of smoking', *J. Am. Med. Assoc.*, **264**, 1541 (1990)

British Medical Association *The BMA Guide to Living with Risk* (London, Penguin, 1990)

Castro F. G. et al 'Cigarette smokers do more than just smoke cigarettes', *Health Psychol.*, **8**, 107 (1989)

Cohen S. & Lichenstein E. 'Perceived stress, quitting smoking, and smoking relapse', *Health Psychol.*, **9**, 466 (1990)

Conway T. L. et al 'Occupational stress and variation in cigarette, coffee, and alcohol consumption', *J. Health Soc. Behav.*, **22**, 155 (1981)

Lee I. M. & Paffenbarger R. S. 'Change in body weight and longevity', *J. Am. Med. Assoc.*, **268**, 2045 (1992)

Leviton A. & Allred E. N. 'Correlates of decaffeinated coffee choice', *Epidemiology*, **5**, 537 (1994)

Lipton R. I. 'Effect of moderate alcohol use on the relationship between stress and depression', *Am. J. Public Health*, **84**, 1913 (1994)

Patterson R. E. et al 'Health lifestyle patterns of US adults', *Prev. Med.*, **23**, 453 (1994)

Perkins K. A. & Grobe J. E. 'Increased desire to smoke during acute stress', *Br. J. Addict.*, **87**, 1037 (1992)

Pohorecky L. A. 'Stress and alcohol interaction', *Alcohol Clin. Exp. Res.*, **15**, 438 (1991)

Strecher V. J. et al 'Do cigarette smokers have unrealistic perceptions of their heart attack, cancer, and stroke risks?' *J. Behav. Med.*, **18**, 45 (1995)

MIND OVER IMMUNE MATTER

Dahlquist G. 'Etiological aspects of insulin-dependent diabetes mellitus', *Autoimmunity*, **15**, 61 (1993)

Dwyer J. *The Body at War. The Story of our Immune System* 2nd edn (London, Dent, 1993)

Irwin M. et al 'Partial sleep deprivation reduces NK cell activity in humans', *Psychosom. Med.*, **56**, 493 (1994)

Rogers M. P. & Fozdar M. 'Psychoneuroimmunology of autoimmune disorders', *Adv. Neuroimmunol.*, **6**, 169 (1996)

Roitt I. M. *Essential Immunology* 8th edn (Oxford, Blackwell, 1994)

Solomon G. F. et al 'Psychoimmunologic and endorphin function in the aged', *Ann. NY Acad. Sci.*, **521**, 43 (1988)

Trinchieri G. 'Biology of NK cells', *Adv. Immunol.*, **47**, 187 (1989)

Weiner H. 'Social and psychobiological factors in autoimmune diseases' in *Psychoneuroimmunology* 2nd edn, ed. R. Ader et al (San Diego, Academic Press, 1991)

THE MIND–IMMUNITY CONNECTIONS

Besedovsky H. O. & Del Rey A. 'Physiological implications of the immune-neuro-endocrine network', in *Psychoneuroimmunology* 2nd edn, ed. R. Ader et al (San Diego, Academic Press, 1991)

Blalock J. E. 'The immune system as a sensory organ', *J. Immunol.*, **132**, 1067 (1984)

Blalock J. E. 'A molecular basis for bidirectional communication between the immune and neuroendocrine systems', *Physiol. Rev.*, **69**, 1 (1989)

Carr D. J. J. & Blalock J. E. 'Neuropeptide hormones and receptors common to the immune and neuroendocrine systems' in *Psychoneuroimmunology* 2nd edn, ed. R. Ader et al (San Diego, Academic Press, 1991)

Felten D. L. et al 'Noradrenergic and peptidergic innervation of lymphoid tissue', *J. Immunol.*, **135**, 755s (1985)

Felten S. Y. & Felten D. L. 'Innervation of lymphoid tissue' in *Psychoneuroimmunology* 2nd edn, ed. R. Ader et al (San Diego, Academic Press, 1991)

Plaut M. 'Lymphocyte hormone receptors', *Ann. Rev. Immunol.*, **5**, 621 (1987)

4 *Mind and Immunity*

GENERAL

Ader R. et al 'Psychoneuroimmunology: interactions between the nervous system and immune system', *Lancet*, **345**, 99 (1995)

Herbert T. B. & Cohen S. 'Stress and immunity in humans', *Psychosom. Med.*, **55**, 364 (1993)

Kiecolt-Glaser J. K. & Glaser R. 'Stress and immune function in humans' in *Psychoneuroimmunology* 2nd edn, ed. R. Ader et al (San Diego, Academic Press, 1991)

O'Leary A. 'Stress, emotion, and human immune function', *Psychol. Bull.*, **108**, 363 (1990)

Stone A. A. & Bovbjerg D. H. 'Stress and humoral immunity', *Adv. Neuroimmunol.*, **4**, 49 (1994)

What can the mind do to the immune system?

Bereavement

Bartrop R. W. et al 'Depressed lymphocyte function after bereavement', *Lancet*, **1**, 834 (1977)

Irwin M. et al 'Impaired NK cell activity during bereavement', *Brain Behav. Immun.*, **1**, 98 (1987)

Kaprio J. et al 'Mortality after bereavement', *Am. J. Health*, **77**, 283 (1987)

Schleifer S. J. et al 'Suppression of lymphocyte stimulation following bereavement', *J. Am. Med. Assoc.*, **250**, 374 (1983)

Zisook S. et al 'Bereavement, depression, and immune function', *Psychiatry Res.*, **52**, 1 (1994)

Nuclear disasters

Collins D. L. & Carvalho A. B. 'Chronic stress from the Goiania ^{137}Cs radiation accident', *Behav. Med.*, **18**, 149 (1993)

Collins D. L. et al 'Coping with chronic stress at Three Mile Island', *Health Psychol.*, **2**, 149 (1983)

Hatch M. C. et al 'Cancer rates after the Three Mile Island nuclear accident', *Am. J. Public Health*, **81**, 719 (1991)

Janerich D. T. 'Can stress cause cancer?' *Am. J. Public Health*, **81**, 687 (1991)

McKinnon W. et al 'Chronic stress, leukocytes, and humoral response to latent viruses', *Health Psychol.*, **8**, 389 (1989)

Pool R. 'A stress-cancer link following accident?' *Nature*, **351**, 429 (1991)

Exams

Dorian B. J. et al 'Aberrations in lymphocytes during psychological stress', *Clin. Exp. Immunol.*, **50**, 132 (1982)

Esterling B. A. et al 'Defensiveness, trait anxiety, and EBV antibody titers in healthy college students', *Health Psychol.*, **12**, 132 (1993)

Glaser R. et al 'Stress-related immune suppression', *Brain Behav. Immun.*, **1**, 7 (1987)

Glaser R. et al 'Stress-associated modulation of proto-oncogene expression in leukocytes', *Behav. Neurosci.*, **107**, 525 (1993)

Glaser R. et al 'Plasma cortisol levels and reactivation of latent EBV in response to exam stress', *Psychoneuroendocrinol.*, **19**, 765 (1994)
Halvorsen R. & Vassend O. 'Effects of examination stress on cellular immunity', *J. Psychosom. Res.*, **31**, 693 (1987)
Jemmott J. B. et al 'Academic stress, power motivation, and decrease in salivary IgA secretion', *Lancet*, **1**, 1400 (1983)
Kiecolt-Glaser J. K. et al 'Modulation of cellular immunity in medical students', *J. Behav. Med.*, **9**, 5 (1986)

Other forms of nastiness

Bachen E. A. et al 'Lymphocyte and cellular immune response to a brief stressor', *Psychosom. Med.*, **54**, 673 (1992)
Bonneau R. H. et al 'Stress-induced effects on cell-mediated response to HSV infection', *Brain Behav. Immun.*, **5**, 274 (1991)
Endresen I. M. et al 'Brief uncontrollable stress and psychological parameters influence human IgM and complement C3', *Behav. Med.*, **17**, 167 (1992)
Esterling B. A. et al 'Chronic stress, social support, and persistent alterations in NK response', *Health Psychol.*, **13**, 291 (1994)
Evans P. et al 'Secretory immunity, mood and life events', *Br. J. Clin. Psychol.*, **32**, 227 (1993)
Farabollini F. et al 'Immune and neuroendocrine response to restraint', *Psychoneuroendocrinol.*, **18**, 175 (1993)
Feng N. et al 'Effect of restraint stress on humoral immune response to influenza virus infection', *Brain Behav. Immun.*, **5**, 370 (1991)
Gerritsen W. et al 'Experimental social fear: immunological, hormonal and autonomic concomitants,' *Psychosom. Med.*, **58**, 273 (1996)
Keller S. E. et al 'Stress-induced changes in immune function in animals' in *Psychoneuroimmunology* 2nd edn, ed. R. Ader et al (San Diego, Academic Press, 1991)
Kiecolt-Glaser J. K. et al 'Spousal caregivers of dementia victims: changes in immunity and health', *Psychosom. Med.*, **53**, 345 (1991)
Kiecolt-Glaser J. K. et al 'Chronic stress alters immune response to influenza vaccine', *Proc. Nat. Acad. Sci. USA*, **93**, 3043 (1996)
Knapp P. H. et al 'Short-term immunologic effects of induced emotion', *Psychosom. Med.*, **52**, 246 (1990)
Meehan R. et al 'The role of psychoneuroendocrine factors on spaceflight-induced immunological alterations', *J. Leukoc. Biol.*, **54**, 236 (1993)
Sheridan J. F. et al 'Restraint stress differentially affects anti-viral immune responses in mice', *J. Neuroimmunol.*, **31**, 245 (1991)
Snyder B. K. et al 'Stress and psychosocial factors: effects on cellular immune response', *J. Behav. Med.*, **16**, 143 (1993)

Stein M. et al 'Influence of brain and behavior on the immune system', *Science*, **191**, 435 (1976)

Zorilla E. P. et al 'Reduced cytokine levels and T-cell function in healthy males: relation to subclinical anxiety', *Brain Behav. Immun.*, **8**, 293 (1994)

Does it matter?

Goodkin K. et al Clinical aspects of psychoneuroimmunology', *Lancet*, **345**, 183 (1995)

Levy S. M. et al 'Persistently low NK cell activity and plasma beta-endorphin: risk factors for infectious disease', *Life Sci.*, **48**, 107 (1991)

Murasko D. M. et al 'Immune reactivity, morbidity, and mortality of elderly humans', *Aging Immunol. Infect. Dis.*, **2**, 171 (1990)

What can the immune system do to the mind?

Anderson J. L. 'The immune system and major depression,' *Adv. Neuroimmunol.*, **6**, 119 (1996)

Cover H. & Irwin M. 'Immunity and depression', *J. Behav. Med.*, **17**, 217 (1994)

Crnic L. S. 'Behavioral consequences of virus infection' in *Psychoneuroimmunology* 2nd edn, ed. R. Ader et al (San Diego, Academic Press, 1991)

Darko D. F. et al 'Plasma beta-endorphin and NK cell activity in major depression', *Psychiatry Res.*, **43**, 111 (1992)

Herbert T. B. & Cohen S. 'Depression and immunity', *Psychol. Bull.*, **113**, 472 (1993)

Husband A. J. 'Role of CNS and behaviour in the immune response', *Vaccine*, **11**, 805 (1993)

Kronfol Z. et al 'Depression, cortisol excretion and lymphocyte function', *Br. J. Psychiatry*, **148**, 70 (1986)

Maes M. 'Evidence for an immune response in major depression', *Progr. Neuropsychopharmacol. Biol. Psychiatry*, **19**, 11 (1995)

Muller N. et al 'Cellular immunity, antigens, and family history of psychiatric disorder in endogenous psychoses', *Psychiatry Res.*, **48**, 201 (1993)

Schiffer R. B. & Hoffman S. A. 'Behavioral sequelae of autoimmune disease' in *Psychoneuroimmunology* 2nd edn, ed. R. Ader et al (San Diego, Academic Press, 1991)

Schleifer S. J. et al 'Major depressive disorder and immunity', *Arch. Gen. Psychiatry*, **46**, 81 (1989)

Shekelle R. B. et al 'Depression and the 17-year risk of death from cancer', *Psychosom. Med.*, **43**, 117 (1981)

Smith A. P. et al 'Selective effects of minor illnesses on human performance', *Br. J. Psychol.*, **78**, 183 (1987)

Stein M. et al 'Depression, the immune system, and health and illness', *Arch. Gen. Psychiatry*, **48**, 171 (1991)
Weisse C. S. 'Depression and immunocompetence', *Psychol. Bull.*, **111**, 475 (1992)

IMMUNE CONDITIONING

Ader R. & Cohen N. 'Behaviorally conditioned immunosuppression and murine systemic lupus erythematosus', *Science*, **215**, 1534 (1982)
Ader R. & Cohen N. 'Psychoneuroimmunology: conditioning and stress', *Ann. Rev. Psychol.*, **44**, 53 (1993)
Ader R. et al 'Conditioned enhancement of antibody production', *Brain Behav. Immun.*, 7, 334 (1993)
Bovbjerg D. et al 'Acquisition and extinction of conditioned suppression of a graft-versus-host response', *J. Immunol.*, **132**, 111 (1984)
Bovbjerg D. H. et al 'Anticipatory immune suppression and nausea in women receiving chemotherapy', *J. Consult. Clin. Psychol.*, **58**, 153 (1990)
Buske-Kirschbaum A. et al 'Conditioned manipulation of NK cells in humans', *Biol. Psychol.*, **38**, 143 (1994)
Cohen N. et al 'Pavlovian conditioning of the immune system', *Int. Arch. Aller. Immunol.*, **105**, 101 (1994)
Fredrikson M. et al 'Trait anxiety and anticipatory immune reactions in women receiving chemotherapy', *Brain Behav. Immun.*, 7, 79 (1993)
Gee A. L. et al 'Behaviorally conditioned modulation of NK cell activity', *Int. J. Neurosci.*, 77, 139 (1994)
Gorczynski R. M. et al 'Tumor growth enhancement in mice demonstrating conditioned immunosuppression', *J. Immunol.*, **134**, 4261 (1985)
Grochowicz P. M. et al 'Behavioral conditioning prolongs heart allograft survival in rats', *Brain Behav. Immun.*, 5, 349 (1991)
Hiramoto R. et al 'Use of conditioning to probe for CNS pathways that regulate fever and NK activity', *Int. J. Neurosci.*, **84**, 229 (1996)
Lysle D. T. et al 'Suppression of adjuvant arthritis by a conditioned aversive stimulus', *Brain Behav. Immun.*, **6**, 64 (1992)
Smith G. R. & McDaniels S. M. 'Psychologically mediated effect on the delayed hypersensitivity reaction to tuberculin', *Psychosom. Med.*, **45**, 65 (1983)
Smith G. R. et al 'Psychologic modulation of the human immune response to varicella zoster', *Arch. Intern. Med.*, **145**, 2110 (1985)
Spector N. H. et al 'Immune enhancement by conditioning of senescent mice', *Ann. NY Acad. Sci.*, **741**, 283 (1994)
Zalcman S. et al 'Immunosuppression elicited by stressors and stressor-related odors', *Brain Behav. Immun.*, **5**, 262 (1991)

THE STRANGE STORY OF THE LEFT-HANDED BRAIN

Aggleton J. P. et al 'Handedness and longevity: archival study of cricketers', *Br. Med. J.*, **309**, 1681 (1994)

Behan P. O. & Geschwind N. 'Hemispheric laterality and immunity' in *Neural Modulation of Immunity* ed. R. Guillemin et al (NY, Raven Press, 1985)

Coren S. 'Handedness and allergic response', *Int. J. Neurosci.*, **76**, 231 (1994)

Coren S. & Halpern D. F. 'Left-handedness: a marker for decreased survival fitness', *Psychol. Bull.*, **109**, 90 (1991)

Fride E. et al 'Immune function in mice selected for high or low degrees of behavioral asymmetry', *Brain Behav. Immun.*, **4**, 129 (1990)

Geschwind N. & Behan P. 'Left-handedness: association with immune disease, migraine, and developmental learning disorder', *Proc. Nat. Acad. Sci. USA*, **79**, 5097 (1982)

Hassler M. & Gupta D. 'Functional brain organisation, handedness and immune vulnerability in musicians', *Neuropsychologia*, **31**, 655 (1993)

Kang D. H. et al 'Frontal brain asymmetry and immune function', *Behav. Neurosci.*, **105**, 860 (1991)

McManus I. C. et al 'Handedness and autoimmune disease', *Lancet*, **341**, 891 (1993)

Tønnessen F. E. et al 'Dyslexia, left-handedness, and immune disorders', *Arch. Neurol.*, **50**, 411 (1993)

Wittling W. & Schweiger E. 'Neuroendocrine brain asymmetry and physical complaints', *Neuropsychologia*, **31**, 591 (1993)

THE WONDERFUL WORLD OF HERPES

Bonneau R. H. et al 'Stress-induced modulation of immune response to HSV infection', *J. Neuroimmunol.*, **42**, 167 (1993)

Corey L. & Spear P. G. 'Infections with HSV', *N. Engl. J. Med.*, **314**, 686 (1986)

Dobbs C. M. et al 'Mechanisms of stress-induced modulation of viral pathogenesis and immunity', *J. Neuroimmunol.*, **48**, 151 (1993)

Gibson J. J. et al 'A cross-sectional study of HSV types 1 and 2 in college students', *J. Infect. Dis.*, **162**, 306 (1990)

Glaser R. & Kiecolt-Glaser J. K. 'Stress-associated depression in cellular immunity: implications for AIDS', *Brain Behav. Immun.*, **1**, 107 (1987)

Glaser R. et al 'Stress-related activation of EBV', *Brain Behav. Immun.*, **5**, 219 (1991)

Kemeny M. E. et al 'Psychological and immunological predictors of genital herpes recurrence', *Psychosom. Med.*, **51**, 195 (1989)

Koff W. C. & Dunegan M. A. 'Neuroendocrine hormones suppress macrophage-mediated lysis of HSV-infected cells', *J. Immunol.*, **136**, 705 (1986)

Kusnecov A. V. et al 'Decreased HSV immunity and enhanced pathogenesis following stressor', *J. Neuroimmunol.*, **38**, 129 (1992)
Longo D. & Koehn K. 'Psychosocial factors and recurrent genital herpes', *Int. J. Psychiatry Med.*, **23**, 99 (1993)
Schmidt D. D. et al 'Stress as a precipitating factor in recurrent herpes labialis', *J. Fam. Pract.*, **20**, 359 (1985)
VanderPlate C. et al 'Genital HSV, stress, and social support', *Health Psychol.*, **7**, 159 (1988)

5 *The Demon Stress*

General

Chrousos G. P. & Gold P. W. 'The concepts of stress and stress system disorders', *J. Am. Med. Assoc.*, **267**, 1244 (1992)
File S. E. 'Recent developments in anxiety, stress and depression', *Pharmacol. Biochem. Behav.*, **54**, 3 (1996)
Sapolsky R. M. *Why Zebras Don't Get Ulcers: a Guide to Stress, Stress-Related Diseases, and Coping* (NY, W. H. Freeman, 1994)
Smith J. C. *Understanding Stress and Coping* (NY, Macmillan, 1993)

What is stress?

Cooper C. L. et al *Living with Stress* (London, Penguin, 1988)
Kamen-Siegel L. et al 'Explanatory style and cell-mediated immunity', *Health Psychol.*, **10**, 229 (1991)
Lechin, F. et al 'Stress versus depression', *Prog. Neuropsychopharmacol. Biol. Psychiat.*, **20**, 899 (1996)
Locke S. E. et al 'Life change stress, psychiatric symptoms, and NK cell activity', *Psychosom. Med.*, **46**, 441 (1984)
Malarkey W. B. et al 'Influence of academic stress on ACTH, cortisol and beta-endorphin', *Psychoneuroendocrinol.*, **20**, 499 (1995)
Peterson C. et al 'Pessimistic explanatory style is a risk factor for physical illness', *J. Pers. Soc. Psychol.*, **55**, 23 (1988)
Ursin H. 'Stress, distress, and immunity', *Ann. NY Acad. Sci.*, **741**, 204 (1994)

The biology of stress

Axelrod J. & Reisine T. D. 'Stress hormones', *Science*, **224**, 452 (1984)
Berk L. S. et al 'Neuroendocrine and stress hormone changes during laughter', *Am. J. Med. Sci.*, **298**, 390 (1989)
Cacioppo J. T. 'Social neuroscience: autonomic, neuro-endocrine and immune responses to stress', *Psychophysiol.*, **31**, 113 (1994)
Fehm-Wolsdorf G. et al 'Auditory reflex thresholds elevated by stress-induced cortisol secretion', *Psychoneuroendocrinol.*, **18**, 579 (1993)

Henkin R. I. 'The effects of corticosteroids and ACTH on sensory organs', *Progr. Brain Res.*, **32**, 270 (1970)
Keller S. E. et al 'Stress-induced suppression of immunity in adrenalectomized rats', *Science*, **221**, 1301 (1983)
Kiecolt-Glaser J. K. et al 'Acute psychological stressors and short-term immune changes', *Psychosom. Med.*, **54**, 680 (1992)
Kirschbaum C. et al 'Stress-induced elevations of cortisol associated with impaired memory', *Life Sci.*, **58**, 1475 (1996)
Manuck S. B. et al 'Prediction of individual differences in cellular immune response', *Psychol. Sci.*, **2**, 111 (1991)
Marotti T. et al 'Met-enkephalin modulates stress-induced alterations of immune responses', *Pharmacol. Biochem. Behav.*, **54**, 277 (1996)
Shavit Y. et al 'Stress, opioid peptides, the immune system, and cancer', *J. Immunol.*, **135**, 834s (1985)
Terman G. W. et al 'Intrinsic mechanisms of pain inhibition: activation by stress', *Science*, **226**, 1270 (1984)

THE QUALITY OF STRESS

Bodnar J. C. & Kiecolt-Glaser J. K. 'Caregiver depression after bereavement', *Psychol. Aging*, **9**, 372 (1994)
Breier A. et al 'Controllable and uncontrollable stress in humans', *Am. J. Psychiatry*, **144**, 1419 (1987)
Keller S. E. et al 'Suppression of immunity by stress', *Science*, **213**, 1397 (1981)
Kort W. J. 'Effect of chronic stress on the immune response', *Adv. Neuroimmunol.*, **4**, 1 (1994)
Laudenslager M. L. et al 'Coping and immunosuppression', *Science*, **221**, 568 (1983)
Monjan A. A. & Collector M. I. 'Stress-induced modulation of the immune response', *Science*, **196**, 307 (1977)
Schedlowski M. et al 'Psychophysiological, neuroendocrine and cellular immune reactions under psychological stress', *Neuropsychobiol.*, **28**, 87 (1993)
Sgoutas-Emch S. A. et al 'Effects of an acute psychological stressor on cardiovascular, endocrine and cellular immune response', *Psychophysiol.*, **31**, 264 (1994)
Sieber W. J. et al 'Modulation of human NK cell activity by exposure to uncontrollable stress', *Brain Behav. Immun.*, **6**, 141 (1992)
Suls J. & Mullen B. 'Life events, perceived control and illness: the role of uncertainty', *J. Hum. Stress*, **7**, 30 (1981)

THE JOY OF STRESS

Dienstbier R. A. 'Arousal and physiological toughness', *Psychol. Rev.*, **96**, 84 (1989)

Hennig J. et al 'Biopsychological changes after bungee jumping', *Neuropsychobiol.*, **29**, 28 (1994)

Rauste-von Wright M. et al 'Psychological characteristics and catecholamine excretion during achievement stress', *Psychophysiol.*, **18**, 362 (1981)

Ursin H. et al *Psychobiology of Stress: a Study of Coping Men* (NY, Academic Press, 1978)

Zeier H. et al 'Effects of work demands on IgA and cortisol in air traffic controllers', *Biol. Psychol.*, **42**, 413 (1996)

6 Other People

GENERAL

Berkman L. F. 'The role of social relationships in health promotion', *Psychosom. Med.*, **57**, 245 (1995)

House J. S. et al 'Social relationships and health', *Science*, **241**, 540. (1988)

Seeman T. E. 'Social ties and health', *Ann. Epidemiol.*, **6**, 442 (1996)

Uchino B. N. et al 'Social support and psysiological processes', *Psychol. Bull.*, **119**, 488 (1996)

HELL IS OTHER PEOPLE? – RELATIONSHIPS AS STRESSORS

Alberts S. C. et al 'Behavioral and immunological correlates of immigration by an aggressive male into a primate group', *Horm. Behav.*, **26**, 167 (1992)

Brown P. C. & Smith T. W. 'Social influence, marriage, and the heart', *Health Psychol.*, **11**, 88 (1992)

Eaker E. D. et al 'Spouse behavior and coronary heart disease in men', *Am. J. Epidemiol.*, **118**, 23 (1983)

Ewart C. K. et al 'High blood pressure and marital discord', *Health Psychol.*, **10**, 155 (1991)

Kiecolt-Glaser J. K. et al 'Negative behavior during marital conflict and immunological down-regulation', *Psychosom. Med.*, **55**, 395 (1993)

Malarkey W. B. et al 'Hostile behavior during marital conflict alters pituitary and adrenal hormones', *Psychosom. Med.*, **56**, 41 (1994)

Manuck S. B. et al 'Social instability and atherosclerosis in monkeys', *Neurosci. Biobehav. Rev.*, **7**, 485 (1983)

Sapolsky R. M. & Mott G. E. 'Social subordinance in wild baboons is associated with suppressed HDL-C', *Endocrinology*, **121**, 1605 (1987)

Stefanski V. & Ben Eliyahu S. 'Social confrontation and tumor metastasis in rats', *Physiol. Behav.*, **60**, 277 (1996)

Swan G. E. et al 'Spouse-pair similarity and husband's coronary heart disease', *Psychosom. Med.*, **48**, 172 (1986)

Weiss R. L. & Aved B. M. 'Marital satisfaction and depression as predictors of physical health', *J. Consult. Clin. Psychol.*, **46**, 1379 (1978)

Hell is alone? – the harmful effects of isolation

Berkman L. & Breslow L. *Health and Ways of Living: Findings from the Alameda County Study* (NY, Oxford Univ. Press, 1983)

Berkman L. F. & Syme S. L. 'Social networks, host resistance, and mortality', *Am. J. Epidemiol.*, **109**, 186 (1979)

Brown G. W. & Harris T. *Social Origins of Depression* (London, Tavistock, 1978)

Bygren L. O. et al 'Attendance at cultural events as determinant for survival', *Br. Med. J.*, **313**, 1577 (1996)

Cohen S. & Syme S. L. (eds.) *Social Support and Health* (NY, Academic Press, 1985)

Collins N. L. et al 'Social support in pregnancy', *J. Pers. Soc. Psychol.*, **65**, 1243 (1993)

Eriksen W. 'The role of social support in the pathogenesis of coronary heart disease', *Fam. Pract.*, **11**, 201 (1994)

Goodwin J. S. et al 'The effect of marital status on treatment and survival of cancer patients', *J. Am. Med. Assoc.*, **158**, 3125 (1987)

Gorkin L. et al 'Psychosocial predictors of mortality', *Am. J. Cardiol.*, **71**, 263 (1993)

Hanson B. S. et al 'Social network and social support influence mortality in elderly men', *Am. J. Epidemiol.*, **130**, 100 (1989)

Jenkinson C. M. et al 'Influence of psychosocial factors on survival after myocardial infarction', *Public Health*, **107**, 305 (1993)

Kaplan G. A. et al 'Social functioning and overall mortality', *Epidemiology*, **5**, 495 (1994)

Kennel J. et al 'Emotional support during labor', *J. Am. Med. Assoc.*, **265**, 2197 (1991)

Orth-Gomer K. & Johnson J. V. 'Social network interaction and mortality', *J. Chron. Dis.*, **40**, 949 (1987)

Orth-Gomer K. et al 'Lack of social support and coronary heart disease', *Psychosom. Med.*, **55**, 37 (1993)

Reynolds P. & Kaplan G. A. 'Social connections and risk for cancer', *Behav. Med.*, **16**, 101 (1990)

Sosa R. et al 'Effect of a supportive companion on perinatal problems', *N. Engl. J. Med.*, **303**, 597 (1980)

Vogt T. et al 'Social networks as predictors of heart disease, cancer, stroke and hypertension', *J. Clin. Epidemiol.*, **45**, 659 (1992)

HOW DOES IT WORK?

Baron R. S. et al 'Social support and immune function among spouses of cancer patients', *J. Pers. Soc. Psychol.*, **59**, 344 (1990)

Broman C. L. 'Social relationships and health-related behavior', *J. Behav. Med.*, **16**, 335 (1993)

Cohen S. et al 'Chronic social stress, affiliation, and immune response in primates', *Psychol. Sci.*, **3**, 301 (1992)

Friedman E. M. et al 'Effects of peer separation on lymphocyte responses in juvenile squirrel monkeys', *Dev. Psychobiol.*, **24**, 159 (1991)

Gerin W. et al 'Social support in social interaction: a moderator of cardiovascular reactivity', *Psychosom. Med.*, **54**, 324 (1992)

Gust D. A. et al 'Effect of a preferred companion in modulating stress in rhesus monkeys', *Physiol. Behav.*, **55**, 681 (1994)

Jemmott J. B. et al 'Motivational syndromes associated with NK cell activity', *J. Behav. Med.*, **13**, 53 (1990)

Kennedy S. et al 'Immunological consequences of stressors: mediating role of interpersonal relationships', *Br. J. Med. Psychol.*, **61**, 77 (1988)

Kiecolt-Glaser J. K. et al 'Cortisol, cellular immunocompetency and loneliness in psychiatric inpatients', *Psychosom. Med.*, **46**, 15 (1984)

Kiecolt-Glaser J. K. et al 'Marital quality, marital disruption, and immune function', *Psychosom. Med.*, **49**, 13 (1987)

Kiecolt-Glaser J. K. et al 'Marital discord and immunity in males', *Psychosom. Med.*, **50**, 213 (1988)

McIntosh W. A. et al 'Life events, social support, and immune response in elderly individuals', *Int. J. Aging Hum. Dev.*, **37**, 23 (1993)

McNaughton M. E. et al 'Stress, social support, and immune status in elderly women', *J. Nerv. Ment. Dis.*, **178**, 460 (1990)

Nerem R. M. et al 'Social environment as a factor in diet-induced atherosclerosis', *Science*, **208**, 1475 (1980)

Serpell J. 'Beneficial effects of pet ownership on human health and behaviour', *J. Roy. Soc. Med.*, **84**, 717 (1991)

Thomas P. D. et al 'Effect of social support on stress-related changes in cholesterol, uric acid and immune function', *Am. J. Psychiatry*, **142**, 735 (1985)

Unden A. L. et al 'Cardiovascular effects of social support in the work place', *Psychosom. Med.*, **53**, 50 (1991)

7 The Wages of Work

General

Bartley M. 'Unemployment and ill health', *J. Epidemiol. Comm. Health*, **48**, 333 (1994)

Fletcher B. *Work, Stress, Disease and Life Expectancy* (Chichester, Wiley, 1991)

Shortt S. E. 'Is unemployment pathogenic?' *Int. J. Health Serv.*, **26**, 569 (1996)

Wilson S. H. & Walker G. M. 'Unemployment and health', *Public Health*, **107**, 153 (1993)

The toad work

Fenwick R. & Tausig M. 'The macroeconomic context of job stress', *J. Health Soc. Behav.*, **35**, 266 (1994)

Frimerman A. et al 'Changes in hemostatic function and occupational stress', *Am. J. Cardiol.*, **79**, 72 (1997)

Haan M. N. 'Job strain and ischaemic heart disease', *Ann. Clin. Res.*, **20**, 143 (1988)

Karasek R. A. et al 'Job characteristics in relation to myocardial infarction', *Am. J. Public Health*, **78**, 910 (1988)

Marmot M. G. et al 'Employment grade and coronary heart disease in civil servants', *J. Epidemiol. Comm. Health*, **32**, 244 (1978)

Pelletier K. & Lutz R. 'Healthy people, healthy business', *Am. J. Health. Prom.*, **2**, 5 (1988)

Schnall P. L. et al 'Job strain, workplace blood pressure and left ventricular mass index', *J. Am. Med. Assoc.*, **263**, 1929 (1990)

Siegrist J. et al 'Low status control, high effort at work and ischemic heart disease', *Soc. Sci. Med.*, **31**, 1127 (1990)

Steptoe A. et al 'Control over work pace, job strain and cardiovascular responses', *J. Hypertens.*, **11**, 751 (1993)

The scourge of unemployment

Arnetz B. B. et al 'Immune function in unemployed women', *Psychosom. Med.*, **49**, 3 (1987)

Beale N. & Nethercott S. 'The nature of unemployment morbidity', *J. Roy. Coll. Gen. Pract.*, **38**, 200 (1988)

Burton P. et al 'Increasing suicide rates among young men in England and Wales', *Br. Med. J.*, **300**, 1695 (1990)

Catalano R. 'The health effects of economic insecurity', *Am. J. Public Health*, **81**, 1148 (1991)

Crombie I. K. 'Can changes in unemployment rates explain recent changes in suicide rates?' *Int. J. Epidemiol.*, **19**, 412 (1990)

Eales M. J. 'Depression and anxiety in unemployed men', *Psychol. Med.*, **18**, 935 (1988)
Ferrie J. E. et al 'Health effects of anticipation of job change and non-employment', *Br. Med. J.*, **311**, 1264 (1995)
Hammarstrom A. 'Health consequences of youth unemployment', *Public Health*, **108**, 403 (1994)
Martikainen P. 'Unemployment and mortality', *Br. Med. J.*, **301**, 407 (1990)
Morris J. K. et al 'Non-employment and changes in smoking, drinking and body weight', *Br. Med. J.*, **304**, 536 (1992)
Moser K. A. et al 'Unemployment and mortality', *Br. Med. J.*, **294**, 86 (1987)
Smith R. '"I'm just not right": the physical health of the unemployed', *Br. Med. J.*, **291**, 1627 (1985)
Smith R. '"We got on each other's nerves": unemployment and the family', *Br. Med. J.*, **291**, 1707 (1985)
Townsend P. et al *Inequalities in Health* (London, Penguin, 1992)
Winefield A. H. et al 'The psychological impact of unemployment and unsatisfactory employment', *Br. J. Psychol.*, **82**, 473 (1991)
Yuen P. & Balarajan R. 'Unemployment and consultation with the general practitioner', *Br. Med. J.*, **298**, 1212 (1989)

8 Sick at Heart

General

Booth-Kewley S. & Friedman H. S. 'Psychological predictors of heart disease', *Psychol. Bull.*, **101**, 343 (1987)
Evans P. D. 'Type A behaviour and coronary heart disease', *Br. J. Psychol.*, **81**, 147 (1990)
Greenwood D. C. et al 'Coronary heart disease: role of psychosocial stress and social support', *J. Public Health Med.*, **18**, 221 (1996)
Johnston D. W. 'The current status of the coronary prone behaviour pattern', *J. Roy. Soc. Med.*, **86**, 406 (1993)

The mind in sudden cardiac death and heart disease

Boltwood M. D. et al 'Anger predicts coronary artery vasomotor response to mental stress', *Am. J. Cardiol.*, **72**, 1361 (1993)
Burg M. M. et al 'Psychological factors in stress-induced silent left ventricular dysfunction', *J. Am. Coll. Cardiol.*, **22**, 440 (1993)
Gottdiener J. S. et al 'Induction of silent myocardial ischemia with mental stress', *J. Am. Coll. Cardiol.*, **24**, 1645 (1994)
Grignani G. et al 'Platelet activation by emotional stress', *Circulation*, **83**, II128 (1991)

Hartel G. 'Psychological factors in cardiac arrhythmias', *Ann. Clin. Res.*, **19**, 104 (1987)

LaVeau P. J. et al 'Transient left ventricular dysfunction during mental stress', *Am. Heart J.*, **118**, 1 (1989)

Rozanski A. et al 'Mental stress and the induction of silent myocardial ischemia', *N. Engl. J. Med.*, **318**, 1005 (1988)

Samuels M. A. 'Neurally induced cardiac damage', *Neurol. Clin.*, **11**, 273 (1993)

Sloan R. P. et al 'Effect of mental stress on cardiac autonomic control', *Biol. Psychol.*, 37, 89 (1994)

Tavazzi L. et al 'Acute pulmonary edema provoked by psychologic stress', *Cardiology*, **74**, 229 (1987)

Williams J. K. et al 'Psychosocial factors impair vascular responses of coronary arteries', *Circulation*, **84**, 2146 (1991)

Williams R. B. & Littman A. B. 'Psychosocial factors: role in cardiac risk and treatment strategies', *Cardiol. Clin.*, **14**, 97 (1996)

Yeung A. C. et al 'Effect of atherosclerosis on the vasomotor response of coronary arteries to mental stress', *N. Engl. J. Med.*, **325**, 1551 (1991)

Coronary-prone personalities and heart disease

Adams S. H. 'Role of hostility in women's health', *Health Psychol.*, **13**, 488 (1994)

Bennett P. & Carroll D. 'Stress management approaches to the prevention of coronary heart disease', *Br. J. Clin. Psychol.*, **29**, 1 (1990)

Cartwright L. K. et al 'What leads to good health in midlife women physicians?' *Psychosom. Med.*, **57**, 284 (1995)

Dembroski T. M. et al 'Components of Type A, hostility and Anger-in: relationships to angiographic findings', *Psychosom. Med.*, **47**, 219 (1985)

Friedman M. & Rosenman R. H. *Type A Behavior and Your Heart* (NY, Knopf, 1974)

Friedman M. et al 'Alteration of Type A behavior and its effects on cardiac recurrences', *Am. Heart J.*, **112**, 653 (1986)

Friedman M. et al 'Effect of Type A behavioral counseling on silent myocardial ischemia', *Am. Heart J.*, **132**, 933 (1996)

Gill J. T. et al 'Reduction of Type A behavior in healthy middle-aged American officers', *Am. Heart J.*, **110**, 503 (1985)

Hardy J. D. & Smith T. W. 'Cynical hostility and vulnerability to disease', *Health Psychol.*, 7, 447 (1988)

Houston B. K. et al 'Behavioral clusters and coronary heart disease risk', *Psychosom. Med.*, **54**, 447 (1992)

Johnston D. W. 'Can and should type A behaviour be changed?' *Postgrad. Med. J.*, **62**, 785 (1986)

Julius M. et al 'Anger-coping types, blood pressure, and all-cause mortality', *Am. J. Epidemiol.*, **124**, 220 (1986)
Markovitz J. H. et al 'Psychological predictors of hypertension', *J. Am. Med. Assoc.*, **270**, 2439 (1993)
Matthews K. A. 'Coronary heart disease and Type A behaviors', *Psychol. Bull.*, **104**, 373 (1988)
O'Brien W. H. & VanEgeren L. 'Perceived susceptibility to heart disease and preventive health behavior', *Behav. Med.*, **17**, 159 (1991)
Roskies E. et al 'The Montreal Type A intervention project', *Health Psychol.*, **5**, 45 (1986)
Siegler I. C. et al 'Hostility during late adolescence predicts coronary risk factors at mid-life', *Am. J. Epidemiol.*, **136**, 146 (1992)
Smith T. W. 'Hostility and health', *Health Psychol.*, **11**, 139 (1992)
Suarez E. C. et al 'Cardiovascular and emotional responses in women: the role of hostility and harassment', *Health Psychol.*, **12**, 459 (1993)
Suls J. & Sanders G. S. 'Type A behavior as a general risk factor for physical disorder', *J. Behav. Med.*, **11**, 201 (1988)
Yakubovich I. S. et al 'Type A behavior pattern and health status after 22 years of follow-up', *Am. J. Epidemiol.*, **128**, 579 (1988)

HOW DOES IT WORK?

Benschop R. J. et al 'Relationships between cardiovascular and immunological changes in stress', *Psychol. Med.*, **25**, 323 (1995)
Evans P. D. & Moran P. 'Cardiovascular unwinding, type A behaviour pattern and locus of control', *Br. J. Med. Psychol.*, **60**, 261 (1987)
Ewart C. K. & Kolodner K. B. 'Predicting ambulatory blood pressure during school', *Psychophysiol.*, **30**, 30 (1993)
Fricchione G. L. et al 'Neuroimmunologic implications in coronary artery disease', *Adv. Neuroimmunol.*, **6**, 131 (1996)
Harbin T. J. 'Type A behavior and physiological responsivity', *Psychophysiol.*, **26**, 110 (1989)
Herbert T. B. et al 'Cardiovascular reactivity and immune response to an acute psychological stressor', *Psychosom. Med.*, **56**, 337 (1994)
Krantz D. S. et al 'Type A behavior and coronary bypass surgery', *Psychosom. Med.*, **44**, 273 (1982)
Matthews K. A. et al 'Cardiovascular reactivity to stress predicts future blood pressure', *Hypertension*, **22**, 479 (1993)
Swan G. E. et al 'Cardiovascular reactivity as a predictor of relapse in smokers', *Health Psychol.*, **12**, 451 (1993)
Vitaliano P. P. et al 'Psychological factors associated with cardiovascular reactivity', *Psychosom. Med.*, **55**, 164 (1993)
Williams R. B. et al 'Type A behavior and elevated physiological and neuroendocrine responses to cognitive tasks', *Science*, **218**, 483 (1982)

9 *The Mind of the Crab*

General

Burgess C. 'Stress and cancer', *Cancer Surv.*, **6**, 403 (1987)

Fife A. et al 'Psychoneuroimmunology and cancer', *Adv. Neuroimmunol.*, **6**, 179 (1996)

Sabbioni M. E. E. 'Psychoneuroimmunological issues in psycho-oncology', *Cancer Invest.*, **11**, 440 (1993)

Spiegel D. 'Psychosocial intervention in cancer', *J. Nat. Cancer Inst.*, **85**, 1198 (1993)

The mind in cancer

Chen C. C. et al 'Adverse life events and breast cancer', *Br. Med. J.*, **311**, 1527 (1995)

Cooper C. L. & Faragher E. B. 'Psychosocial stress and breast cancer', *Psychol. Med.*, **23**, 653 (1993)

Flach J. & Seachrist L. 'Mind–body meld may boost immunity', *J. Nat. Cancer Inst.*, **86**, 256 (1994)

Fox B. H. 'Depressive symptoms and risk of cancer', *J. Am. Med. Assoc.*, **262**, 1231 (1989)

Hilakivi-Clarke L. et al 'Psychosocial factors in breast cancer', *Breast Cancer Res. Treat.*, **29**, 141 (1994)

Horne R. L. & Picard R. S. 'Psychosocial risk factors for lung cancer', *Psychosom. Med.*, **41**, 503 (1979)

Kune G. A. et al 'Personality as a risk factor in large bowel cancer', *Psychol. Med.*, **21**, 29 (1991)

Persky V. W. et al 'Personality and risk of cancer', *Psychosom. Med.*, **49**, 435 (1987)

Riley V. 'Psychoneuroendocrine influences on immunocompetence and neoplasia', *Science*, **212**, 1100 (1981)

Shaffer J. et al 'Family attitudes in youth as a possible precursor of cancer', *J. Behav. Med.*, **5**, 143 (1982)

Sklar L. S. & Anisman H. 'Stress and coping factors influence tumor growth', *Science*, **205**, 513 (1979)

Visintainer M. A. et al 'Tumor rejection in rats after inescapable or escapable shock', *Science*, **216**, 437 (1982)

Is there a cancer-prone personality?

Eysenck H. J. 'Psychosocial factors, cancer, and ischaemic heart disease', *Br. Med. J.*, **305**, 457 (1992)

Fox C. M. et al 'Loneliness, emotional repression, marital quality, and life events in breast cancer', *J. Comm. Health*, **19**, 467 (1994)

Gross J. 'Emotional expression in cancer onset and progression', *Soc. Sci. Med.*, **28**, 1239 (1989)
Jasmin C. et al 'Evidence for a link between psychological factors and breast cancer', *Ann. Oncol.*, **1**, 22 (1990)
Quander-Blaznik J. 'Personality as a predictor of lung cancer', *Pers. Indiv. Diff.*, **12**, 125 (1991)
Shaffer J. W. et al 'Personality traits in youth and subsequent cancer', *J. Behav. Med.*, **10**, 441 (1987)
Temoshok L. 'Personality, coping style, emotion and cancer', *Cancer Surv.*, **6**, 545 (1987)
Temoshok L. & Dreher H. *The Type C Connection* (NY, Random House, 1992)
Wirsching M. et al 'Psychological identification of breast cancer patients before biopsy', *J. Psychosom. Res.*, **26**, 1 (1982)

PSYCHOLOGICAL INFLUENCES ON SURVIVAL

Andrykowski M. A. et al 'Psychosocial factors predictive of survival after bone marrow transplantation', *Psychosom. Med.*, **56**, 432 (1994)
Fallowfield L. 'Psychosocial interventions in cancer', *Br. Med. J.*, **311**, 1316 (1995)
Fawzy F. I. et al 'Psychosocial interventions in cancer care', *Arch. Gen. Psychiatry*, **52**, 100 (1995)
Greer S. et al 'Psychological response to breast cancer and 15-year outcome', *Lancet*, **335**, 49 (1990)
Hislop T. G. et al 'Psychosocial factors in breast cancer', *J. Chronic Dis.*, **40**, 729 (1987)
Meyer T. J. & Mark M. M. 'Effects of psychosocial interventions with cancer patients', *Health Psychol.*, **14**, 101 (1995)
Pettingale K. W. et al 'Mental attitudes to cancer: an additional prognostic factor', *Lancet*, **1**, 750 (1985)
Ramirez A. J. et al 'Stress and relapse of breast cancer', *Br. Med. J.*, **298**, 291 (1989)
Spiegel D. et al 'Effect of psychosocial treatment on survival of patients with metastatic breast cancer', *Lancet*, **2**, 888 (1989)
Spiegel D. 'Psychosocial aspects of breast cancer treatment', *Semin. Oncol.*, **24** (S1), 36 (1997)
Stavraky K. M. et al 'The effect of psychosocial factors on lung cancer mortality', *J. Clin. Epidemiol.*, **41**, 75 (1988)

HOW DOES IT WORK?

Andersen B. L. 'Surviving cancer', *Cancer*, **74**, 1484 (1994)
Ayres A. et al 'Influence of mood and adjustment on compliance with chemotherapy', *J. Psychosom. Res.*, **38**, 393 (1994)

Ben-Eliyahu S. et al 'Stress increases metastatic spread of a mammary tumor in rats', *Brain Behav. Immun.*, **5**, 193 (1991)
Bovbjerg D. H. & Valdimarsdottir H. 'Familial cancer, emotional distress, and low natural cytotoxic activity in healthy women', *Ann. Oncol.*, **4**, 745 (1993)
Chaitchik S. & Kreitler S. 'Induced versus spontaneous attendance of breast-screening tests', *J. Cancer Educ.*, **6**, 43 (1991)
Glaser R. et al 'Effects of stress on an important DNA repair mechanism', *Health Psychol.*, **4**, 403 (1985)
Harris J. R. et al 'Breast cancer', *N. Engl. J. Med.*, **327**, 319 (1992)
Keinan G. et al 'Predicting women's delay in seeking medical care after discovery of a lump in the breast', *Behav. Med.*, **17**, 177 (1991)
Kiecolt-Glaser J. K. et al 'Distress and DNA repair in human lymphocytes', *J. Behav. Med.*, **8**, 311 (1985)
Kreitler S. et al 'The psychological profile of women attending breast-screening tests', *Soc. Sci. Med.*, **31**, 1177 (1990)
Levy S. et al 'Correlation of stress with sustained depression of NK cell activity and prognosis in breast cancer', *J. Clin. Oncol.*, **5**, 348 (1987)
Petitto J. M. et al 'Genetic differences in social behavior: relation to NK function and tumor development', *Neuropsychopharmacol.*, **8**, 35 (1993)
Romero L. et al 'Possible mechanism by which stress accelerates growth of virally-derived tumors', *Proc. Nat. Acad. Sci. USA*, **89**, 11084 (1992)

10 Encumbered with Remedies

General

Ader R. 'On the clinical relevance of psychoneuroimmunology', *Clin. Immunol. Immunopathol.*, **64**, 6 (1992)
Buckman R. & Sabbagh K. *Magic or Medicine?* (London, Macmillan, 1993)
Hall N. R. S. & O'Grady M. P. 'Psychosocial interventions and immune function' in *Psychoneuroimmunology* 2nd edn, ed. R. Ader et al (San Diego, Academic Press, 1991)

Relax!

Burnette M. M. et al 'Control of genital herpes recurrences using progressive muscle relaxation', *Behav. Ther.*, **22**, 237 (1991)
Decker T. W. et al 'Relaxation therapy as an adjunct in radiation oncology', *J. Clin. Psychol.*, **48**, 388 (1992)
Green M. L. et al 'Daily relaxation modifies immunoglobulins and symptom severity', *Biofeedback Self-Regul.*, **13**, 187 (1988)
Jasnoski M. L. & Kugler J. 'Relaxation, imagery, and neuroimmunomodulation', *Ann. NY Acad. Sci.*, **496**, 722 (1987)

Kiecolt-Glaser J. K. et al 'Psychosocial enhancement of immunocompetence in a geriatric population', *Health Psychol.*, **4**, 25 (1985)
McGrady A. et al 'The effects of biofeedback-assisted relaxation on immunity, cortisol, and white blood cell count', *J. Behav. Med.*, **15**, 343 (1992)
Morse D. R. et al 'Stress induced sudden cardiac death: can it be prevented?' *Stress Medicine*, **8**, 35 (1992)
Van Rood Y. et al 'The effects of stress and relaxation on the immune response', *J. Behav. Med.*, **16**, 163 (1992)

EXERCISE!

Blumenthal J. A. et al 'Aerobic exercise reduces responses to mental stress', *Am. J. Cardiol.*, **65**, 93 (1990)
Brahmi Z. et al 'The effect of acute exercise on NK cell activity', *J. Clin. Immunol.*, **5**, 321 (1985)
Crews D. J. & Landers D. M. 'Aerobic fitness and reactivity to psychosocial stressors', *Med. Sci. Sports Exerc.*, **19**, 114 (1987)
Eaton C. B. et al 'Physical activity predicts long-term coronary heart disease and all-cause mortalities', *Arch. Fam. Med.*, **4**, 323 (1995)
Farmer M. E. et al 'Physical activity and depressive symptoms', *Am. J. Epidemiol.*, **128**, 1340 (1988)
Gerhardsson M. et al 'Sedentary jobs and colon cancer', *Am. J. Epidemiol.*, **123**, 775 (1986)
Grossarth-Maticek R. et al 'Sport activity and personality in preventing cancer and coronary heart disease', *Percept. Mot. Skills*, **71**, 199 (1990)
Hoffman-Goetz L. 'Exercise, natural immunity, and tumor metastasis', *Med. Sci. Sports Exerc.*, **26**, 157 (1994)
Lee I. M. et al 'Exercise intensity and longevity in men', *J. Am. Med. Assoc.*, **273**, 1179 (1995)
Mackinnon L. T. 'Exercise immunology', *Med. Sci. Sports Exerc.*, **26**, 191 (1994)
Nieman D. C. 'Exercise, infection, and immunity', *Int. J. Sports Med.*, **15**, S131 (1994)
Pedersen B. K. & Bruunsgaard H. 'How physical exercise influences the establishment of infection', *Sports Med.*, **19**, 393 (1995)
Roth D. L. & Holmes D. S. 'Influence of physical fitness on the impact of stressful life events', *Psychosom. Med.*, **47**, 164 (1985)
Shephard R. J. & Shek P. N. 'Exercise, aging and immune function', *Int. J. Sports Med.*, **16**, 1 (1995)
Simon H. B. 'Exercise and human immune function' in *Psychoneuroimmunology* 2nd edn, ed. R. Ader et al (San Diego, Academic Press, 1991)

Psychoneuroimmunology and AIDS

Antoni M. H. et al 'Stress management buffers distress and immunologic changes following notification of HIV-1 seropositivity', *J. Consult. Clin. Psychol.*, **59**, 906 (1991)

Cohen S. I. 'Voodoo death, the stress response, and AIDS', *Adv. Biochem. Psychopharmacol.*, **44**, 95 (1988)

Evans D. L. et al 'Stress-associated reductions of cytotoxic T lymphocytes and NK cells in asymptomatic HIV infection', *Am. J. Psychiatry*, **152**, 543 (1995)

Goodkin K. et al 'Psychoneuroimmunology and HIV-1 infection revisited', *Arch. Gen. Psychiatry*, **51**, 246 (1994)

Hassan N. F. & Douglas S. D. 'Stress-related neuroimmunomodulation in HIV-1 infection', *Clin. Immunol. Immunopathol.*, **54**, 220 (1990)

Jewett J. F. & Hecht F. M. 'Preventive health care for adults with HIV infection', *J. Am. Med. Assoc.*, **269**, 1144 (1993)

Kemeny M. E. 'Psychoneuroimmunology of HIV infection', *Psychiat. Clin. North Am.*, **17**, 55 (1994)

Kessler R. C. et al 'Stressful life events and symptom onset in HIV-1 infection', *Am. J. Psychiatry*, **148**, 733 (1991)

Nott K. H. et al 'Psychology, immunology, and HIV', *Psychoneuroendocrinol.*, **20**, 451 (1995)

Patterson T. L. et al 'Stress and depressive symptoms predict immune change among HIV-positive men', *Psychiatry*, **58**, 299 (1995)

Perry S. et al 'Lymphocyte subsets and psychosocial variables among adults with HIV', *Arch. Gen. Psychiatry*, **49**, 396 (1992)

Solomon G. F. et al 'Psychoneuroimmunologic aspects of HIV' in *Psychoneuroimmunology* 2nd edn, ed. R. Ader et al (San Diego, Academic Press, 1991)

Temoshok L. R. 'HIV/AIDS, psychoneuroimmunology and beyond', *Adv. Neuroimmunol.*, **3**, 141 (1993)

Imagery, miracle cures and other exotica

Baider L. et al 'Progressive muscle relaxation and guided imagery in cancer patients', *Gen. Hosp. Psychiatry*, **16**, 340 (1994)

Dowling St J. 'Lourdes cures and their medical assessment', *J. Roy. Soc. Med.*, **77**, 634 (1984)

Gruber B. L. et al 'Immunological responses of breast cancer patients to behavioral interventions', *Biofeedback Self-Regul.*, **18**, 1 (1993)

Hall H. et al 'Changes in neutrophil adherence following imagery', *Int. J. Neurosci.*, **85**, 185 (1996)

Lutgendorf S. K. et al 'Cognitive coping strategies predict EBV-antibody titre following stressor disclosure', *J. Psychosom. Res.*, **38**, 63 (1994)

Pennebaker J. W. et al 'Disclosure of traumas and immune function', *J. Consult. Clin. Psychol.*, **56**, 239 (1988)

Petrie K. J. et al 'Disclosure of trauma and immune response to hepatitis B vaccination', *J. Consult. Clin. Psychol.*, **63**, 787 (1995)

Rider M. S. & Achterberg J. 'Effect of music-assisted imagery on neutrophils and lymphocytes', *Biofeedback Self-Regul.*, **14**, 247 (1989)

Siegel B. S. *Peace, Love and Healing: Body-Mind Communication and the Path to Self-Healing* (NY, Harper & Row, 1989)

Simonton O. C. et al *Getting Well Again: a Step-By-Step Self-Help Guide to Overcoming Cancer* (LA, Tarcher, 1978)

Zachariae R. et al 'Effect of relaxation and guided imagery on cellular immune function', *Psychother. Psychosom.*, **54**, 32 (1990)

KILL OR CURE?

Grossarth-Maticek R. & Eysenck H. J. 'Prophylactic effects of psychoanalysis', *J. Behav. Ther. Exp. Psychiatry*, **21**, 91 (1990)

Horowitz M. J. et al 'The stressful impact of news of risk for premature heart disease', *Psychosom. Med.*, **45**, 31 (1983)

Rostrup M. & Ekeberg O. 'Awareness of high blood pressure influences on psychological and sympathetic responses', *J. Psychosom. Res.*, **36**, 117 (1992)

Stoate H. G. 'Can health screening damage your health?' *J. Roy. Coll. Gen. Pract.*, **39**, 193 (1989)

11 *Exorcising the Ghost in the Machine*

Black S. *Mind and Body* (London, William Kimber, 1969)

Bunge M. *The Mind–Body Problem* (Oxford, Pergamon Press, 1980)

Carter R. B. *Descartes' Medical Philosophy* (Baltimore, Johns Hopkins Univ. Press, 1983)

Cottingham J. *Descartes* (Oxford, Blackwell, 1986)

Damasio A. R. *Descartes' Error* (NY, Grosset/Putnam, 1994)

Edelman G. M. *Bright Air, Brilliant Fire* (NY, Basic Books, 1992)

Entralgo P. L. *Mind and Body* (London, Harvill Press, 1955)

Humphrey N. *A History of the Mind* (London, Chatto & Windus, 1992)

Ostenfeld E. *Ancient Greek Psychology and the Modern Mind-Body Debate* (Aarhus, Aarhus Univ. Press, 1987)

Wilson M. D. *Descartes* (London, Routledge & Kegan Paul, 1978)

12 *A Fresh Pair of Lenses*

GENERAL

Nesse R. M. & Williams G. C. *Evolution and Healing. The New Science of Darwinian Medicine* (London, Weidenfeld & Nicolson, 1995)

DEVELOPMENT

Ackerman S. H. et al 'Premature maternal separation and lymphocyte function', *Brain Behav. Immun.*, **2**, 161 (1988)

Boyce W. T. et al 'Temperament and the psychobiology of childhood stress', *Pediatrics*, **90**, 483 (1992)

Brunner E. et al 'Childhood social circumstances and plasma fibrinogen', *Lancet*, **347**, 1008 (1996)

Coe C. L. et al 'Early rearing conditions alter immune responses in the developing infant primate', *Pediatrics*, **90**, 505 (1992)

De Jonge F. H. et al 'Developmental aspects of social stress in pigs', *Physiol. Behav.*, **60**, 389 (1996)

Gehde E. & Baltrusch H. J. F. 'Early experience and development of cancer in later life', *Int. J. Neurosci.*, **51**, 257 (1990)

Hessing M. J. et al 'Individual behavioral and physiological strategies in pigs', *Physiol. Behav.*, **55**, 39 (1994)

Jemerin J. M. & Boyce W. T. 'Psychobiological differences in childhood stress response', *J. Dev. Behav. Pediatr.*, **11**, 140 (1990)

Lau R. R. et al 'Development and change of young adults' preventive health beliefs and behavior', *J. Health Soc. Behav.*, **31**, 240 (1990)

Laudenslager M. et al 'Behavioral and immunological consequences of brief mother-infant separation', *Dev. Psychobiol.*, **23**, 247 (1990)

Lewis M. 'Individual differences in response to stress', *Pediatrics*, **90**, 487 (1992)

Lubach G. R. et al 'Effects of early rearing environment on immune responses of infant rhesus monkeys', *Brain Behav. Immun.*, **9**, 31 (1995)

Musante L. et al 'Consistency of children's hemodynamic responses to laboratory stressors', *Int. J. Psychophysiol.*, **17**, 65 (1994)

O'Grady M. P. & Hall N. R. S. 'Long-term effects of neuroendocrine-immune interactions during early development' in *Psychoneuroimmunology* 2nd edn, ed. R. Ader et al (San Diego, Academic Press, 1991)

Sallis J. F. et al 'Blood pressure reactivity in children', *J. Psychosom. Res.*, **32**, 1 (1988)

von Hoersten S. et al 'Effect of early experience on behavior and immune response', *Physiol. Behav.*, **54**, 931 (1993)

Evolution

Dawkins R. *The Selfish Gene* (Oxford, Oxford Univ. Press, 1976; new edn, 1989)

Dawkins R. *River out of Eden* (London, Weidenfeld & Nicolson, 1995)

Doran T. F. et al 'Acetaminophen: more harm than good for chicken pox?' *J. Pediatr.*, **114**, 1045 (1989)

Goldsmith M. F. 'Ancestors may provide clinical answers', *J. Am. Med. Assoc.*, **269**, 1477 (1993)

Graham N. M. et al 'Adverse effects of aspirin, acetaminophen and ibuprofen in rhinovirus-infected volunteers', *J. Infect. Dis.*, **162**, 1277 (1990)

Leakey R. *The Origin of Humankind* (London, Weidenfeld & Nicolson, 1994)

Munck A. & Guyre P. M. 'Glucocorticoids and immune function' in *Psychoneuroimmunology* 2nd edn, ed. R. Ader et al (San Diego, Academic Press, 1991)

Ottaviani E. & Franceschi C. 'Neuroimmunology of stress from invertebrates to man', *Prog. Neurobiol.*, **48**, 421 (1996)

Profet M. 'The function of allergy: immunological defense against toxins', *Q. Rev. Biol.*, **66**, 23 (1991)

Profet M. 'Pregnancy sickness as adaptation' in *The Adapted Mind* ed. J. H. Barkow et al (NY, Oxford Univ. Press, 1992)

Williams G. C. & Nesse R. M. 'The dawn of Darwinian medicine', *Q. Rev. Biol.*, **66**, 1 (1991)

Darwin's illness

Bernstein R. E. 'Darwin's illness: Chagas' disease resurgens' *J. Roy. Soc. Med.*, **77**, 608 (1984)

Bowlby J. *Charles Darwin* (London, Hutchinson, 1990)

INDEX